Residential
Equipment Selection

ACCA
Air Conditioning Contractors of America

Manual S

Acknowledgments

Manual S is a comprehensive guide for selecting and sizing residential heating and cooling equipment.

The development and publication of this manual was commissioned by the **Air Conditioning Contractors of America (ACCA).** Much of the information in this manual was secured from, or based on, information published in the **American Society of Heating, Refrigerating and Air-Conditioning Engineers (ASHRAE)** handbooks, and from various design manuals and engineering data sheets published by equipment manufacturers.

Grateful acknowledgment is made to those members of the **Manual S** review committee who contributed to the preparation of this manual.

This manual was written by **Hank Rutkowski,** P.E., ACCA technical director.

ACCA Manual S Review Committee:

Dick Pasterkamp — Chairman
Pasterkamp Htg & A/C Co.
1930 S. Cherokee
Denver, CO 80223

Glenn Friedman
Engineered Air Systems
720 12th St.
Richmond, CA 94801

Joe Boito
Penelec
1001 Broad Street
Johnstown, PA 15907

Jim Herritage
Energy Auditors Inc.
1161 Park Way Dr.
Mt. Pleasant, SC 29464

Pat Jopek
Merit Mechanical Systems Inc.
18W371 N. Frontage Road
Darien, IL 60561

Bert Magnuson
Jack Frost Co. Inc.
2112 109th Street South
Tacoma, WA 98444

Bud Healy
Director of Education
NHARW
1389 Dublin Road
Columbus, OH 43215

©1995, Air Conditioning Contractors of America
ISBN: 1-892765-03-9

Reader Response

ACCA is dedicated to providing its members and users of all ACCA manuals with accurate, up-to-date and useful information. If you believe that any of the information contained in this manual is incomplete or inaccurate, if you have suggestions for improving the manual, or if you have general comments, we'd like to hear from you. Please write your comments below and return this form to:

Air Conditioning Contractors of America
1712 New Hampshire Avenue, N.W.
Washington, D.C. 20009
Phone: 202/483-9370 • Fax: 202/588-1217

Comments on ACCA's **Manual S**:

_____

_____

_____

_____

_____

_____

Name _____

Company _____

Address _____

_____

Phone _____

Introduction

This manual documents the procedures that should be used to select and size residential cooling equipment, furnaces and heat pumps. These procedures emphasize the importance of using performance data that documents the sensible, latent or heating capacity for a wide variety of operating conditions. This manual also suggests sizing strategies for all types of cooling and heating equipment and it discusses the nuances of the presentation formats that are used by equipment manufacturers.

All readers are encouraged to study Sections 1 and 8. Section 1 is a prerequisite because it provides an overview of the procedures and considerations that affect equipment selection. Section 8 is useful because it explains why certification ratings should not be used for selecting equipment. The other sections of this manual, which discuss the sizing procedures that apply to particular types of equipment, can be read on an as-needed basis. This information covers:

Furnaces and boilers

Cooling-only equipment

Air-to-air heat pumps

Water-to-air heat pumps

Dual fuel heat pumps

Multispeed and variable-speed equipment

Also note the useful information that is contained in the appendix sections.

Additional information about residential comfort conditioning equipment can be found in companion publications. Refer to **Manual H** for a comprehensive discussion of air-to-air, water-to-air and dual-fuel heat pump equipment. Also note that ACCA has a publication that discusses buried water-loop piping systems (design and sizing issues) and a "Residential Systems" manual that deals with comfort, air quality, zoning and system selection issues.

Table of Contents

Section 1

Basic Concepts

Section 2

Furnaces and Boilers

Section 3

Cooling Equipment Selection Procedure

Section 3 (Continued)

Cooling Equipment Selection Procedure

Section 4

Air-Source Heat Pump

Section 5

Water-to-Air Heat Pump

Section 5 (Continued)

Water-to-Air Heat Pump

Section 6

Dual Fuel Systems

Section 7

Variable and Dual Speed Equipment

Section 7 (Continued)

Variable- and Dual-speed Equipment

Section 8

Limitations of the ARI Certification Data

Appendix 1

Summary of Tables and Equations

Appendix 2

Manufacturers Application Data

Appendix 3

Performance Models

Appendix 4

Furnace Cycling Efficiency

Appendix 5

DX Coil Matching

Appendix 6

Effect of Altitude

Appendix 7

Air-source Heat Pump Supply Temperatures

Appendix 7 (Continued)

Air-source Heat Pump Supply Temperatures

Section 1
Basic Concepts

1-1 Load Calculations

An accurate load analysis must be performed before the equipment can be selected. Use ACCA **Manual J** and **Form J-1** to calculate the design heating load, the sensible cooling load and the latent cooling load.

Use the local outdoor temperature when selecting air-cooled condensing equipment. Refer to **Manual J**, Table-1, (2-1/2 percent design dry bulb column).

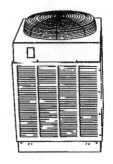

Air-Cooled Condensing Units

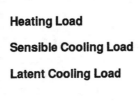

Heating Load

Sensible Cooling Load

Latent Cooling Load

Use the ground water temperature or the earth-coupled water loop temperature (late summer) when selecting water cooled equipment.

1-2 Operating Conditions for Cooling Equipment

Water-Cooled Condensing Equipment

The capacity of direct expansion (DX) cooling equipment, which could be an air conditioner or heat pump, will be affected either by the temperature of the outdoor air (when air-cooled equipment is installed) or by the temperature of the condenser water (when water-cooled equipment is installed). (The equipment capacity of water-cooled equipment also is affected by the water-side flow rate.) Therefore, when cooling equipment is selected, the output capacity will depend on the operating conditions that are expected to occur at the building site.

Cooling cycle performance also depends on the dry-bulb (db) and wet-bulb (wb) temperatures of the air that enters the indoor coil. If the design does not include mechanical ventilation, and if no significant conduction or leakage losses are associated with the return-side of the duct system, these temperatures will be approximately equal to the indoor temperatures. (The condition of the indoor air is defined by the **Manual J** indoor design temperature and desired relative humidity value.)

- The outdoor design temperature for sizing air-cooled air conditioners and heat pumps is normally equal to the **Manual J** 2-1/2 percent dry-bulb temperature. However, if the condensing unit is located on a roof, the air temperature may be 10 °F to 20 °F greater than the **Manual J** value.

- The design temperature for sizing water-source equipment may either equal the local ground water temperature or the temperature of the water that is contained in an earth-coupled (closed loop) piping system. (Note that the ground water temperature is fairly constant throughout the whole year, but during late summer, the temperature of the water that is circulating in an earth-coupled piping loop can be more than 90 °F.)

With no outdoor air and no return-side duct gains, the condition of the indoor air and the condition of the entering air are the same.

Entering dry-bulb = 75°F
Entering wet-bulb = 63°F
(50 % relative humidity)

Indoor Coil with No Ventilation

Note that the entering wet-bulb temperature depends on the relative humidity of the return air. When there is no mechanical ventilation, the condition of the air entering the indoor coil can be estimated by using Table 1-1.

Entering Conditions — Cooling Coil No Mechanical Ventilation, No Return-Side Duct Gains		
Relative Humidity	Entering db	Entering wb
55 Percent	75	64
50 Percent	75	62-1/2 (Use 63)
45 Percent	75	61-1/2 (Use 62)

Table 1-1

If mechanical ventilation is provided (by a return-side intake), the dry-bulb temperature of the air entering the indoor coil will be a little higher than the room temperature, and in most places, the entering wet-bulb will also be a little higher than the room wet-bulb temperature. (In arid climates mechanical ventilation will reduce the entering wet-bulb temperature.)

With ventilation and no return-side duct gains, the condition of the indoor air and the condition of the entering air are not the same.

Entering dry-bulb exceeds 75°F
Entering wet-bulb depends on the local climate (usually exceeds 64 wb)

Indoor Coil With Return-side Ventilation

Usually, the mechanical ventilation rate is less than 10 percent of the supply cubic feet per minute (CFM) value. At a 10 percent ventilation rate, the condition of the air that approaches the indoor coil is (approximately) represented by the dry-bulb and wet-bulb values that are listed in Table 1-2.

Approximate Entering Conditions — Cooling Coil 10 Percent Outdoor Air, No Return-Side Duct Gains				
Relative Humidity	Humid Climate		Dry Climate	
	Entering db	Entering wb	Entering db	Entering wb
55 Percent	77	65-1/2	77	64
50 Percent	77	64	77	62-1/2
45 Percent	77	62-1/2	77	61
Refer to **Manual P**, for a procedure that can be used to precisely determine the condition of a mixture of return air and outdoor air.				

Table 1-2

In Figure 1-2, the dry climate dry-bulb and wet-bulb values represent the condition of the entering air when dry outdoor air is mixed with indoor air that has a relative humidity that ranges from 45 percent to 55 percent. However, in some cases, the indoor humidity could be lower than 45 percent if the internal moisture gains cannot balance the moisture losses that are associated with infiltration and ventilation. If this is the case, the entering wet-bulb will be less than 61 °F and the indoor coil could be completely dry. In other words, if the climate is very dry, a humidification device will be required to maintain the indoor humidity at 45 percent or higher.

1-3 Operating Conditions for Heat Pump Heating

The heating capacity of air-source heat pumps is sensitive to the temperature of the air that enters the outdoor air-to-refrigerant coil. The heating capacity of water-source heat pumps is sensitive to the temperature of the water that enters the water-to-refrigerant heat exchanger. (The capacity of water source equipment also is affected by the water-side flow rate.) Therefore, when heat pump equipment is selected, the output capacity will depend on the operating conditions that are expected to occur at the building site.

* During the heating season, the design temperature for sizing air-cooled heat pumps is equal to the **Manual J** outdoor design temperature.

* The design temperature for sizing water-source equipment may be equal to the local ground water temperature or the temperature of the water that is contained in an earth-coupled (closed loop) system. (As previously noted, the ground water temperature is fairly constant throughout the whole year, but the temperature of the water that is contained in an earth-coupled piping loop can fall below the freezing point during late winter.)

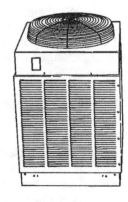

Use the local outdoor temperature when selecting air-source heat pump equipment. Refer to **Manual J**, Table-1, (97-1/2 percent design dry-bulb column).

Air Source Heat Pump Equipment

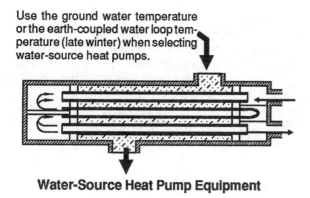

Use the ground water temperature or the earth-coupled water loop temperature (late winter) when selecting water-source heat pumps.

Water-Source Heat Pump Equipment

Heat pump heating performance also is affected by the dry-bulb temperature of the air entering the indoor refrigerant coil. If no ventilation air is involved, the design value for the entering air temperature will be approximately equal to the **Manual J** room design temperature, which is 70 °F.

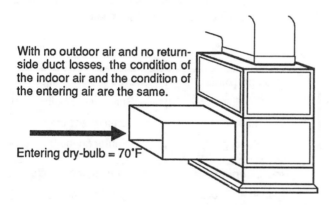

With no outdoor air and no return-side duct losses, the condition of the indoor air and the condition of the entering air are the same.

Entering dry-bulb = 70°F

Indoor Heat Pump Unit with No Ventilation

If outdoor air is used for mechanical ventilation, the entering air temperature may be 2 to 8 degrees colder than the indoor design temperature. At a 10 percent ventilation rate, the temperature of the air that approaches the indoor coil is equal to the values that are listed in Table 1-3.

Approximate Entering Conditions — Indoor Coil 10 Percent Outdoor Air, 70 °F Return Air							
Outdoor db	-10	0	10	20	30	40	50
Entering db	62	63	64	65	66	67	68

Table 1-3

1-4 Operating Conditions, Forced-Air Furnaces

Furnace capacity is not affected by the outdoor temperature and it is not greatly affected by the entering dry-bulb temper-

ature. However, furnaces are subject to temperature rise limitations, which are discussed in Section 2.

1-5 Estimating the Cooling CFM

When a cooling system is required, an approximate blower CFM value can be obtained by using the **Manual J** cooling load information to estimate the desired temperature difference between the room air and the supply air. Once this temperature difference is known, the sensible heat equation can be used to obtain the cooling CFM value. The details of this procedure are summarized below.

Step 1
Use the **Manual J** design loads (**Form J-1**, Procedure D) to calculate the coil sensible heat ratio (SHR). This sensible heat ratio is equal to the **Manual J** sensible load divided by the **Manual J** total load.

$$SHR = \frac{Manual\ J\ Sensible\ Load}{Manual\ J\ Total\ Load}$$

For example, if the sensible load is equal to 30,000 BTUH and the latent load is equal to 5,000 BTUH, the total load is equal to 35,000 BTUH and the sensible heat ratio is 0.86. This means that 86 percent of the cooling load is a sensible load and that 14 percent of the cooling load is a latent load.

$$SHR = \frac{30000}{(30000 + 5000)} = 0.86$$

Step 2
Determine the most desirable temperature for the supply air. This temperature will depend on the relative size of the sensible and latent loads, as indicated by the sensible heat ratio.

- If the sensible heat ratio is relatively high, the latent load is relatively small compared to the sensible load, so the indoor humidity can be controlled with a coil that is a little warmer. (Leaving-air temperature should be about 58 °F.)

- If the sensible heat ratio is relatively low, the latent load is relatively large compared to the sensible load, so satisfactory humidity control requires a cold coil. (Leaving-air temperature air should be about 54 °F.)

On the next page, Table 1-4 summarizes the relationship between the sensible heat ratio and the temperature of the air leaving the coil (LAT). The table also provides a value for the desired room-air-supply-air temperature difference (TD).

Table 1-4 demonstrates that when the sensible heat ratio is high (SHR above 0.85), the dry-bulb temperature of the

Sensible Heat Ratio Versus TD Value			
SHR	LAT	Room db	TD
Below 0.80	54	75	21
0.80 to 0.85	56	75	19
Above 0.85	58	75	17

Table 1-4

leaving air can be moderated without loosing control over the indoor humidity. In this case, the TD value can be about 17 °F. Conversely, when the sensible heat ratio is low (SHR below 0.80), the coil (and the dry-bulb temperature of the leaving air) has to be colder. In this case a TD value of about 21 °F will be required to control the indoor humidity.

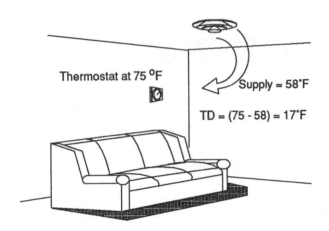

Thermostat at 75 °F

Supply = 58˚F

TD = (75 - 58) = 17˚F

Supply Air Temperature Difference

Step 3
After the temperature difference value is determined, use the sensible heat equation (provided below) to obtain an approximate value for the cooling CFM. (In this equation, the 1.1 is an "air properties" constant that is associated with a sea-level elevation, but this value is normally used for elevations that are lower than 1,500 feet. At 5,000 feet the value of this constant is reduced to about 0.92, because the air is less dense at altitude. Refer to Appendix 6 for more information about altitude corrections.)

$$CFM = \frac{Manual\ J\ Sensible\ Load}{1.1 \times TD}$$

For example, if the sensible load is equal to 32,000 BTUH and the TD value is equal to 21 °F (humid climate), the desired air flow through the coil will be about 1,385 CFM. Or, if the sensible load is equal to 32,000 BTUH and the TD value is equal to 19 °F (low outdoor humidity), the desired air flow through the coil will be about 1,531 CFM.

Note that the cooling CFM depends only on the sensible load and the TD value. Less CFM is required if colder supply air is used, and more CFM is required if warmer supply air is used. However, if the supply air is too warm, it may not be dry enough to absorb the latent load. Therefore, the system designer is not free to arbitrarily set the design TD value. If Table 1-4 is used to select the TD value, the supply CFM will have enough sensible capacity to neutralize the sensible load and enough latent capacity to absorb the latent load.

1-6 Equipment Selection Data Summary

The designer should summarize the equipment performance requirements before referring to the manufacturer's performance data. This summary should include the **Manual J** loads; the outdoor design conditions; the indoor design conditions; the condition of the air entering the indoor coil (which is the same as the indoor design condition when there is no return-side ventilation or return-side duct leakage); the estimated cooling CFM; and, if water-source equipment is involved, the entering water temperature. Table 1-5 summarizes the information that is required for equipment sizing.

Design Loads	Outdoor Conditions
Sensible	Summer dry-bulb
Latent	Summer wet-bulb
Heating	Winter dry-bulb
Room Conditions	**Air Entering Indoor Coil**
Dry-bulb — Cooling	Dry-bulb — cooling
Relative humidity	Wet-bulb — cooling
Dry-bulb — heating	Dry-bulb — heating
Air Flow Estimate	**Water Temperature**
TD (from Table 1-4)	Late summer
CFM (from equation)	Late winter

Table 1-5

1-7 Comprehensive Cooling Data Required

After the equipment selection parameters are collected, the manufacturer's performance data can be used to select a piece of equipment. This data consists of a series of tables that are organized according to capacity. (The progression is normally from smaller to larger.) Carefully search the information sheets for the unit that provides the desired performance characteristics. (There may be more than one unit that can satisfy the performance requirements without being unacceptably oversized or undersized.)

Wet Bulb (°F)	Air Vol. (CFM)	Total Cool Cap. (BTUH)	Comp. Motor Watts Input	Sensible to Total Ratio S/T (Dry Bulb)			Total Cool Cap. (BTUH)	Comp. Motor Watts Input	Sensible to Total Ratio S/T (Dry Bulb)			Total Cool Cap. (BTUH)	Comp. Motor Watts Input	Sensible to Total Ratio S/T (Dry Bulb)		
		85					**95**					**105**				
				76	80	84			76	80	84			76	80	84
63	1,000	29,900	2,820	.76	.88	.97	28,300	3,000	.78	.90	.97	26,700	3,190	.80	.93	.97
	1,125	30,600	2,840	.79	.92	.97	28,900	3,020	.82	.95	.97	27,200	3,210	.84	.97	.97
	1,250	31,200	2,850	.82	.96	.97	29,400	3,040	.85	.97	.97	27,900	3,250	.88	.97	.97
67	1,000	31,800	2,870	.59	.71	.82	30,100	3,070	.61	.73	.84	28,300	3,270	.62	.75	.87
	1,125	32,400	2,890	.61	.74	.86	30,500	3,080	.63	.76	.88	28,700	3,290	.65	.78	.91
	1,250	32,800	2,900	.63	.77	.89	30,900	3,100	.65	.79	.92	29,100	3,300	.67	.82	.96

Note: All values are gross capacities and do not include indoor blower coil motor heat deduction.

Table 1-6

Note that each manufacturer publishes equipment performance data in a different format. It usually takes a little time to familiarize yourself with the data published by a particular manufacturer.

An example of a comprehensive cooling performance data sheet is provided by Table 1-6. Notice that this data accounts for all the parameters that affect the performance of air-source equipment (CFM, outdoor temperature, entering air dry-bulb and entering air wet-bulb). Also note that the data sheet provides information on total capacity, sensible capacity and latent capacity. (In this case the sensible capacity equals the total capacity multiplied by the S/T ratio.) Also note that performance parameters are tabulated for a full range of operating conditions:

- Low, medium, and high blower speeds
- Four outdoor temperatures (115 °F not shown)
- Three entering dry-bulb temperatures
- Three entering wet-bulb temperatures (71 °F not shown)

Because different manufacturers publish performance data in different formats, it is always necessary to study the information sheets and the footnotes that supplement the data tables. For example, consider Table 1-7. Notice that the table lists values for total capacity (TC) and sensible capacity (SHC). Also notice that these values are tabulated for three fan speeds, a range of outdoor temperatures and a range of entering wet-bulb temperatures. However, this table does not account for the effect of the entering dry-bulb temperature. But an inspection of the footnote reveals that the data is correct if the

Temp (F) Air Entering Outdoor Unit		**Air Entering Indoor Unit - CFM / BF**							
		970 / .11			**1080 / .11**			**1200 / .12**	
		Indoor Unit Entering Air Temperature - Ewb (F)							
		72	67	62	72	67	62	72	67
85	TC	33.0	30.3	27.4	33.2	30.6	27.9	33.2	30.9
	SHC	17.0	21.8	25.8	17.4	22.8	26.8	17.8	23.8
	Kw	4.32	4.17	3.99	4.38	4.23	4.07	4.44	4.30
95	TC	31.8	28.8	25.8	32.0	29.0	26.2	32.1	29.3
	SHC	16.6	21.3	24.8	17.1	22.3	25.7	17.7	23.4
	Kw	4.58	4.38	4.20	4.64	4.45	4.26	4.71	4.54
100	TC	30.9	27.8	25.0	31.1	28.0	25.4	31.3	28.4
	SHC	16.3	21.0	24.3	16.9	21.8	25.0	17.5	23.1
	Kw	4.70	4.48	4.28	4.76	4.54	4.37	4.84	4.63

Dry-bulb temperature of air entering coil equals 80 °F.
Indoor blower heat has been deducted from cooling capacity values.

Table 1-7

entering dry-bulb is equal to 80 °F; therefore, a sensible capacity correction is required if the entering dry-bulb temperature is not equal to 80 °F.

Table 1-8 provides an example of this manufacturer's solution to this problem. In this case a second dry-bulb correction factor table supplements the primary table. (This two-table approach is an older format that may be out of use. Refer to Appendix 2 for examples of a single-table format.)

Bypass Factor (BF)	Entering Air Dry-Bulb Temperature (°F)					
	79	78	77	76	75	Under 75
	81	82	83	84	85	Over 85
	Correction Factor					
.10	.98	1.96	2.94	3.92	4.91	Use formula shown below
.20	.87	1.74	2.62	3.49	4.36	
.30	.76	1.53	2.29	3.05	3.82	
Entering db Correction						
Below 80 °F, subtract (corr. x fact.) from SHC value						
Above 80 °F, add (corr. x fact.) to SHC value						

Table 1-8

1-8 Heat Pump — Heating Performance Data

Table 1-9 provides an example of a comprehensive heat pump heating capacity table. Notice that this table provides information on the heating output for the full range of winter design temperatures. Also notice that the footnote that accompanies the table reassures the designer that the heating capacity values include an allowance for reverse operation during the defrost cycle. (The term "integrated capacity" is often used when the heating output is adjusted to account for the defrost cycle.)

Once again, it is important to read the fine print. Notice that the footnote states that the capacities listed in the main body of the table correspond to a 70 °F entering dry-bulb tempera-

ture and that a correction is required if the entering dry-bulb is not equal to 70 °F. However, Table 1-10 indicates that this correction amounts only to a small percentage of the listed capacity value, so it can be ignored with little consequence.

Temperature of Air Entering Indoor Coil (°F)	Correction Factors	
	Capacity	Power
65	1.02	0.99
70	1.00	1.00
75	0.98	1.01

Table 1-10

1-9 Blower Performance Data

The blower performance data also should be included in the manufacturer's literature. In the example provided by Table 1-11, the CFM and the corresponding external static pressure is tabulated at three fan speeds (high, medium, and low).

Unit	CFM	Fan Speed		
		High	Med	Low
		External Static Pressure		
HP - 24	1250	0.80	0.60	0.49
	1300	0.74	0.54	0.38
	1450	0.68	0.46	0.26
	1500	0.59	0.35	—
	1650	0.50	—	—
HP - 30	1450	0.85	0.75	0.65
	1550	0.73	0.60	0.45
	1650	0.58	0.42	0.18
	1750	0.41	0.22	—
	1850	0.23	—	—
Static pressure values do not include an allowance for a wet coil or for electric heating coils.				

Table 1-11

Outdoor Unit	Indoor Unit	Integrated Heating Capacities*													
		Temperature of Air Entering Outdoor Unit (db)												
		-10		0		10		17		20		30		40	
		Cap	Kw	Cap	Kw	Cap	Kw	Cap	Kw	Cap	Kw	Cap	Kw	Cap	Kw
015	018	4.5	1.3	5.5	1.6	7.0	1.6	8.5	1.7	9.0	1.7	11.0	1.7	13.5	1.8
020	024	5.0	1.3	7.0	1.5	8.6	1.6	10.0	1.7	10.5	1.7	13.0	1.7	15.5	1.8
027	024	6.0	2.2	8.5	2.4	11.5	2.5	13.5	2.6	14.0	2.6	17.5	2.7	21.0	2.9
* Integrated capacity reflects a reduction in heating capacity caused by defrost cycles. Air entering indoor coil at 70 °F.															

Table 1-9

Again, it is important to read the footnotes. Notice that the static pressure data has allowed for the pressure drop across a dry coil and for the pressure drop across a standard throwaway filter. However, this particular static pressure data table does not include an allowance for the additional pressure drop that occurs when the coil is wet or for the pressure drop across a supplemental electric resistance heating coil.

In this case, the manufacturer will provide a table that lists the additional pressure drops associated with undocumented operating conditions and accessory devices (see Table 1-12). When the duct system is designed, these pressure drops will have to be subtracted from the external static pressure value that is listed in the blower performance table.

Component Pressure Drops (IWC)			
Unit	CFM	Wet DX Coil	Resistance Heater
HP - 24	950	0.05	0.08
	1050	0.07	0.10
	1150	0.08	0.12
	1250	0.09	0.13

Table 1-12

1-10 ARI Certification Data

The performance data that appears in the Air Conditioning and Refrigeration Institute (ARI) **Certified Products Directory** is intended only to provide a basis for comparing the performance of various makes and models of cooling units and heat pumps. This information is not equivalent to the manufacture's application data.

Cooling Data
Some air conditioning and heat pump manufacturers do not publish comprehensive cooling performance data. Unfortunately the ARI certification data does not provide all the information that is necessary for selecting cooling equipment.

• The sensible capacity is required to select cooling equipment, but the ARI data provides information only on total capacity.

• The latent capacity of the cooling equipment also must be checked, but the ARI data provides information only on total capacity.

• During cooling, direct expansion coil performance is sensitive to the CFM across the coil. Therefore, performance data that corresponds to the available fan speeds is required. However, the ARI data does not provide this information. (The sensible capacity at the allowable maximum or minimum CFM could be 5 to 10 percent more or less than the sensible capacity at the midrange CFM.)

• The duct sizing calculations require knowledge of the supply CFM, but the ARI data does not include a CFM value.

• The duct sizing calculations require knowledge of the available external static pressure that is produced by the blower, but the ARI directory does not provide blower information.

Also note that the information in the ARI directory is not compatible with the **Manual J** indoor design conditions and it may not be compatible with the local outdoor design temperature or the local ground water temperature.

• The most commonly used design value for the indoor dry-bulb is equal to the 75 °F value that is recommended by **Manual J**, but the ARI data is only valid when the indoor dry-bulb is equal to 80 °F. This is important because the sensible capacity of a unit is affected by the temperature of the air entering the indoor coil.

• The most commonly used value for the indoor wet-bulb temperature is 63 °F or 64 °F. (This corresponds to a room that is maintained at 75 °F dry-bulb and 50 or 55 percent relative humidity.) However, the ARI data specifies performance only at a 67 °F wet-bulb temperature. (This corresponds to 80 °F dry-bulb and 50 percent relative humidity.) This is important because the sensible and latent capacity of a unit is affected significantly by the wet-bulb temperature of the air entering the indoor coil.

• The summer design temperature for most locations ranges from 85 °F to 115 °F. To select air-source equipment accurately, the performance data must be compatible with the local design temperature. However, the ARI data documents performance only for an outdoor temperature of 95 °F.

• If water-source equipment is used, the cooling data must be compatible with the local ground water temperature or the closed water loop temperature. In any particular instance, the source water temperature can range from less than 40 °F to more than 90 °F. The ARI data provides performance data only for water temperatures of 50 °F and 70 °F.

Heating Data
Some heat pump manufacturers do not publish comprehensive heating capacity data. In this case, the ARI certification data is available, but it does not provide all the information that is necessary for selecting heat pump equipment.

• Air-source heat pump sizing calculations require heating performance data that corresponds to the local winter design temperature. Most locations have a winter design temperature that falls between -20 °F and 50 °F. However, the ARI data documents heat pump performance only at the 47 °F rating point.

- Air-source heat pump calculations should include a balance-point diagram. This diagram cannot be produced with the ARI data because the heating performance is documented only at the 47 °F rating point.

- When the outdoor temperature is between 10 °F and 40 °F the "reverse cycle" operation that occurs during the defrost cycle reduces the net heating capacity of an air source heat pump by 5 to 10 percent. This capacity reduction should be shown on the balance-point diagram. The ARI data does not provide any information about this penalty.

- If water-source equipment is used, the heating data must be compatible with the local ground water temperature or the closed water loop temperature. In any particular instance, the source water temperature can range from less than 30 °F to more than 70 °F. The ARI data provides performance data only for water temperatures of 50 °F and 70 °F.

Section 2
Furnaces and Boilers

2-1 Design Heating Load

An accurate load analysis must be performed before selecting the heating equipment. Use **Manual J** and the **J-1 Form** to calculate the design heating load. If a forced-air furnace is involved, the sizing load should include the duct losses. If a hot water boiler is being selected, the sizing load should include the piping losses. However, it is not necessary to add an exceptionally large startup load or a setback recovery load to the **Manual J** heating load.

- System startup operations (commissioning a new system or putting a system online after a service call) are exceptional events and they do not always occur during the most severe weather. But, even if a "design load" startup is required, the indoor temperature will moderate within an hour or so, a marginal level of comfort will be experienced within 2 to 6 hours and uncompromised comfort will eventually be restored — even if the system has no excess heating capacity.

- When recovery from a night setback is involved, it will take 10 to 20 minutes to route all of the air that is contained within the conditioned space through the furnace. Even if a furnace is oversized by 100 percent, the cycle time for this process will be reduced only by 5 to 10 minutes.

- The time that is takes to heat the floor, wall, and ceiling panels, and the furnishings, is not reduced in proportion to the amount of excess furnace capacity because the convective heat transfer that occurs at these points is not affected by the size of the furnace.

- When recovery from a night setback is involved, the time that is takes to reach an acceptable threshold level of comfort is considerably shorter than the time it takes for the furnace to cycle off. (People begin to feel comfortable in about one-third to one-half of the time that it takes to reheat the thermal mass that is associated with the indoor surfaces and the furniture.)

- Since oversized equipment is not desirable, a programmable thermostat can (and should) be used to activate the heating equipment before the occupants begin their day.

2-2 Sizing Guidelines

Oversizing is not recommended because comfort may be compromised when a furnace or boiler short-cycles. (This may be less of a problem when variable speed equipment is involved.) Ideally, the output capacity of the furnace or boiler

must be greater than the design heating load, but not more than 40 percent larger than the design heating load.

However, if year-round comfort is desired, and if the cooling load is large in comparison to the heating load, a significantly oversized furnace may be required to obtain blower performance that is compatible with the size of the cooling coil. (Heat pump systems, electric resistance coils and domestic water-heater heating coils may be preferable to a furnace or a boiler when a home has a large cooling load and a small heating load.)

Oversizing is not an issue as far as fuel efficiency is concerned. Laboratory tests sponsored by the U.S. Department of Energy indicate that modern furnaces and boilers can be oversized by as much as 100 percent without causing a significant increase in the operating cost. (Refer to Appendix 4 for more information.)

Table 2-1 provides a summary of the sizing guidelines that are associated with residential heating equipment. These guidelines relate the amount of excess heating capacity to heating season comfort, air-side performance during the cooling season, and operating cost.

Limits on Excess Heating Capacity		
Criterion	Furnace	Boiler
Heating Comfort	40 %	40 %
Cooling Comfort	May exceed 40%	NA
Heating Efficiency	100 %	100 %

Table 2-1

2-3 Manufacturer's Performance Data

Always use the manufacturer's performance data to select a forced-air furnace or a boiler. At the top of the next page, Table 2-2 provides an example of the required heating performance data. Always use the output capacity value to size the heating equipment.

For example, if the **Manual J** heat loss is equal to 56,000 BTUH, the furnace should have an output capacity that falls between 56,000 BTUH and 78,400 BTUH. Table 2-2 shows that an H-80 furnace has an output capacity that falls within the desired range.

H-80 Heating Performance (BTUH) Heating (H) and Heating-Cooling (HC) Models				
Model	50	60	80	100
Input BTUH	50,000	60,000	80,000	100,000
Output BUH	39,000	47,000	62,000	78,000
AFUE	80%	80%	80%	80%
Temp. Rise	25-55	25-55	35-65	35-65

Table 2-2

2-4 Blower Performance — Heating Only

When a heating-only system is installed, the required blower pressure is considerably smaller than the pressure associated with a year-round system because the supply air does not have to pass through a cooling coil. And, if cooling is not required, the output heating capacity can be delivered by any flow rate that is compatible with the furnace heat exchanger temperature rise limitations (see section 2-6). This "loose" set of pressure-airflow requirements makes it relatively easy to find a furnace that has adequate blower performance.

For example, Table 2-3 summarizes the blower data for the model H-80 furnace. This figure shows that, depending on the fan speed, the flow through the furnace will range from 690 to 1,360 CFM when the flow resistance that is produced by the air distribution system varies between 0.6 and 0.2 inches water column (IWC). In other words, if the temperature rise is acceptable, the duct system can be designed for any flow rate that falls between 690 and 1,360 CFM. (Refer to **Manual D** for information about residential duct sizing procedures.)

H-80 Blower Performance Air Flow (CFM) versus Fan Speed					
Fan Speed	External Resistance (IWC)				
	0.20	0.30	0.40	0.50	0.60
Low	770	770	760	735	690
Med-low	935	925	905	670	830
Med-high	1,200	1,180	1,145	1,105	1,045
High	1,360	1,315	1,265	1,195	1,125
Output Capacity = 62,000 BTUH					

Table 2-3

2-5 Blower Performance — Heating and Cooling

When a year-round comfort system is installed, the required blower pressure is significantly larger than the pressure asso-ciated with a heating-only system because the supply air must pass through a cooling coil. (Coil pressure drops range from about 0.15 IWC to more than 0.25 IWC.) In addition, the blower must produce a flow rate (CFM value) that is compatible with the refrigeration equipment and the cooling loads (sensible and latent). Normally, this flow rate is higher than the flow rate that would be required for a heating-only system.

* A specific amount of sensible cooling usually requires more air flow than the same amount of furnace heating (cooling load equals heating load, for example) because the temperature drop across the cooling coil typically is smaller than the temperature rise across the furnace heat exchanger. (Cooling coils usually operate with a tempera-ture drop of 17 °F to 21 °F, depending on the size of the latent load; but the temperature rise across furnace heat exchangers can vary from about 25 °F to more than 45 °F.)

* Contemporary construction practices tend to reduce the differential between the heating load and the cooling load.

* Small heating loads and large cooling loads are associated with mild and warm climates.

Therefore, if the cooling load is not significantly smaller than the heating load, the air distribution system must be designed to accommodate the air flow rate that is required for the cooling season. This means that a year-round conditioning system may require a furnace that has a more powerful blower than would be required for a heating-only system.

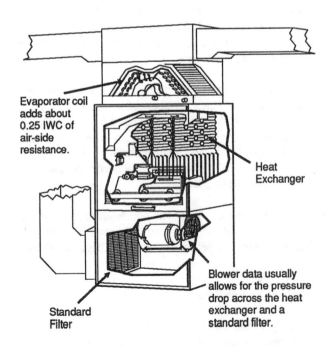

Evaporator coil adds about 0.25 IWC of air-side resistance.

Heat Exchanger

Standard Filter

Blower data usually allows for the pressure drop across the heat exchanger and a standard filter.

How Air-side Devices Dissipate Pressure

For example, assume that the **Manual J** heating load is equal to 56,000 BTUH and that 1,400 CFM is required for cooling. Table 2-2 shows that the H-80 furnace is the correct size as far as capacity is concerned, but Table 2-3 shows that it will not deliver the desired air flow. (A flow rate of 1,360 CFM is acceptable, but the resistance produced by the refrigerant coil and the air distribution system will certainly exceed the 0.2 external resistance value associated with the high blower speed option.) In this case it will be necessary to search for a furnace that has the same heating capacity and a more powerful blower.

Since year-round comfort systems are so common, manufactures are producing low-output furnaces that feature a relatively powerful blower. Some manufacturers even produce two versions of the same furnace — a heating-only model and a heating-cooling model. In this case, both models will have the same heating capacity, but the year-round model will be equipped with a more powerful blower.

In regard to the previous example, Table 2-4 indicates that the manufacturer's heating-cooling model (HC-80) has the correct amount of heating capacity and the desired blower performance. (A half inch of external static pressure normally will be enough to move the air through the cooling coil and the air distribution system, provided that the duct runs are sized according to procedures presented in **Manual D**.)

Forced-air Furnace Sizing Procedure	
Year-round Applications	
Step 1	Select a furnace that satisfies the heating load without exceeding the 40 percent limitation on excess output capacity.
Step 2	Compare the blower performance of the furnace with the required cooling CFM. The blower must be able to deliver the desired cooling CFM (within 10 percent) when it is operating against an external resistance of approximately 0.5 inches water column. (Adequate blower performance at a midrange fan speed is preferable, but not mandatory.)
Step 3	If the blower performance is inadequate, check the air delivery of other products that have the correct heating capacity.
Step 4	If there are no units that have the correct combination of heating capacity and blower performance, ignore the 40 percent oversizing rule and select the smallest furnace that will provide the desired cooling season air flow rate.
Step 5	If this procedure leads to a furnace that is grossly oversized, consider using a different type of heating system.

Table 2-5

2-6 Temperature Rise

After the furnace is selected, be sure to check the temperature rise across the heat exchanger. The calculated rise must be within the range specified in the manufacturer's performance data. The sensible heat equation can be used for this work. (Sometimes, the rise associated with the cooling CFM value may be too small. When this is the case, a slower fan speed can be used during the heating season.)

$$Rise \ (°F) = \frac{Output \ Capacity \ (BTUH)}{1.1 \ x \ Heating \ CFM}$$

For example, if 1,400 CFM is required for cooling and the output capacity of the HC-80 furnace is equal to 62,000 BTUH, the rise across the heat exchanger will be equal to about 40 °F. This value falls within the range specified by the furnace manufacturer (see Table 2-2).

$$Rise = \frac{62000 \ BTUH}{1.1 \ x \ 1400} = 40 \ °F$$

HC-80 Blower Performance					
Air Flow (CFM) Versus Fan Speed					
Fan Speed	**External Resistance (IWC)**				
	0.20	**0.30**	**0.40**	**0.50**	**0.60**
Low	1,010	1,035	1,045	1,045	1,030
Med-Low	1,300	1,305	1,295	1,275	1,240
Med-High	1,470	1,460	1,440	1,410	1,370
High	1,875	1,805	1,735	1,660	1,580
Output Capacity = 62,000 BTUH					

Table 2-4

If equipment manufacturers do not offer a product that has the desired combination of heating capacity and blower performance, the size of the furnace will be dictated by the cooling season air flow requirement. In this case, the output heating capacity may exceed the 40 percent oversizing rule. (If the excess heating capacity exceeds the 40 percent guideline by a significant amount, the designer should consider using another type of year-round conditioning system — a heat pump system, for example.) Table 2-5 provides a summary of the forced-air furnace sizing procedure.

2-7 Other Considerations

The location of the primary heating equipment must be compatible with the floor plan of the home and the space that is available for installing the distribution and venting systems.

If a forced-air furnace is featured, the relationship between the location of the duct system and the location of the furnace will dictate the desired equipment arrangement (up flow, down flow or horizontal). Be sure the furnace has the required fuel efficiency rating (AFUE) if a state code, local code, or utility regulation mandates a minimum level of efficiency. Also make sure that the fuel train controls and combustion controls comply with all codes and regulations, that the furnace has an adequate supply of combustion air (during any possible operating condition), and that the venting system complies with all building codes, fuel gas codes, and utility regulations. (*Note that as far as venting is concerned, the likelihood of a condensation problem increases in proportion to the amount of excess heating capacity.*)

Section 3
Cooling Equipment Selection Procedures

3-1 Basic Types of Cooling Equipment

The sizing procedures outlined in this section apply to central cooling-only equipment, central air-to-air heat pumps, and water-to-air heat pump equipment. This machinery either could be packaged in one cabinet (single-package unit) or could be provided in a split configuration. (Split installations include the cooling-only hardware that is used with forced-air furnaces, which consist of an outdoor condensing unit and a direct expansion coil.)

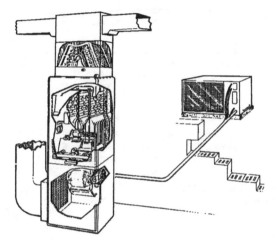

Furnace with Cooling Coil

3-2 Manufacturer's Application Data Required

Comprehensive performance data is required for the equipment selection process. This data should document the sensible and latent capacity of the equipment for a wide range of operating conditions. This information normally is published by manufacturers who provide "total system packages." However, comprehensive performance data is not available for mixed and matched components pieced together by a third party. (Refer to Appendix 5 for information about matching condensing units with blower coils.)

> Some manufacturers do not publish comprehensive equipment performance data, but it may be available upon request. If the appropriate capacity information is not available, the designer is advised to use a product that is supported by the necessary documentation. (Refer to Section 8 of this manual for an explanation of why the ARI rating data should not be used for equipment selection.)

3-3 Design Loads

Equipment selection should be based on the **Manual J** cooling loads for the entire house. Both the sensible and the latent loads are required for this work.

3-4 Sizing Limitations

Cooling equipment should be sized to satisfy the **Manual J** design loads (sensible and latent) when the system is operating at the summer design condition. (The summer design condition is defined by the indoor and outdoor conditions that were used to produce the **Manual J** load estimate.) This rule applies to all types of air-to-air and water-to-air equipment.

Gross oversizing is not recommended because part-load temperature control, humidity control, operating costs, and installation costs are adversely affected when the equipment has an excessive amount of cooling capacity. In this regard, the acceptable amount of excess capacity will depend on the application.

- Cooling-only equipment should be sized so that the total cooling capacity does not exceed the total cooling load by more than 15 percent.

- If heat pump equipment (air-source or water-source) is installed in a warm or moderate climate, the total cooling capacity should not exceed the total cooling load by more than 15 percent.

- If heat pump equipment (air-source or water-source) is installed in a cold climate (where heating costs are a primary concern), the total cooling capacity can exceed the total cooling load by as much as 25 percent. (This allows the designer to place more emphasis on refrigeration-cycle heating performance).

3-5 Cooling Capacity Is Conditional

The capacity of direct expansion cooling equipment is affected by four parameters that define the operating condition. On the next page, Figure 3-1 shows that the total, sensible, and latent cooling capacity of a unit is affected by variations in the air flow across the indoor coil, by changes in the wet-bulb temperature of the air entering the indoor coil, by changes in the dry-bulb temperature of the air entering the indoor coil, and by changes in the temperature of the fluid flowing through the condenser.

Air flow Across the Coil

As the air flow (CFM) through the indoor coil increases, the total capacity and the sensible capacity increase. However, the latent capacity will decrease because the rate of increase in sensible capacity is greater than the rate of increase in total capacity.

Condensing Temperature

As the condensing temperature—outdoor air temperature (OAT) or entering water temperature (EWT)—increases, the total capacity, sensible capacity, and latent capacity will all decrease. In this case, the latent capacity decreases because the rate of decrease in the sensible capacity is smaller than the rate of decrease in the total capacity.

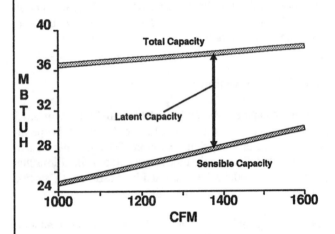

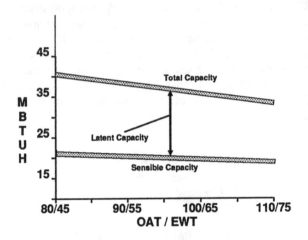

Entering Wet-bulb Temperature

As the wet-bulb temperature of the air entering the indoor coil increases, the total capacity increases and the sensible capacity decreases. In this case, the latent capacity will increase as the difference between the total capacity and the sensible capacity increases.

Entering Dry-bulb Temperature

As the dry-bulb temperature of the air entering the indoor coil increases, the total capacity stays relatively constant and the sensible capacity increases. In this case, the latent capacity decreases because the difference between the total capacity and the sensible capacity gets smaller as the sensible capacity increases.

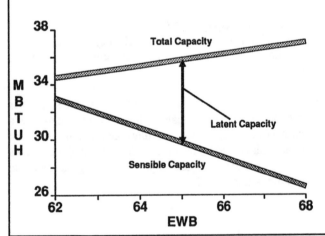

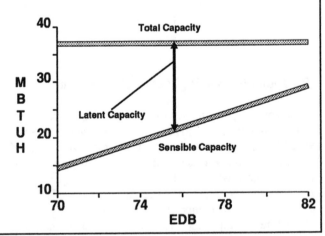

Figure 3-1

3-6 Excess Latent Capacity

Sometimes, the performance data indicates that the latent capacity of cooling equipment exceeds the estimated latent load by a few thousand BTUH. If this equipment is installed in the home, the indoor humidity will be a little lower than the value that was used to calculate the latent load. This, in turn, will produce a slight decrease in total capacity, a significant decrease in latent capacity, and a significant increase in sensible capacity (see Entering Wet-bulb Temperature, Figure 3-1). In other words, the imbalance between the latent capacity and the latent load tends to be self-correcting.

- On a hot summer day, the relative humidity in the conditioned space will be a little lower than the relative humidity value that was used for the **Manual J** calculation, but even if it drops by 5 percent, the indoor conditions still will be very comfortable.

- Since the entering wet-bulb temperature is lowered, about half of the excess latent capacity will be converted to sensible capacity. This converted capacity can be added to the sensible capacity value that was extracted from the manufacturer's performance data.

3-7 Mechanical Ventilation

If mechanical ventilation is required— and if the outdoor air CFM does not exceed 10 percent of the blower CFM— the equipment sizing procedure is basically the same as the procedure that would be used if no ventilation were required. The only difference between these two cases is the condition of the air that enters the indoor coil. This difference is summarized by Tables 1-1 and 1-2, which are reproduced below. (In Table 1-2, the dry climate wet-bulb values are based on the assumption that there is enough moisture generation within the home to produce a small latent load on the coil. But if the climate is extremely dry, this will not be the case; the indoor humidity will be less than 45 percent and the entering wet-bulb temperature will be less than 61 °F.)

Entering Conditions — Cooling Coil No Mechanical Ventilation, No Return-side Duct Gains		
Relative Humidity	Entering db	Entering wb
55 Percent	75	64
50 Percent	75	62-1/2 (Use 63)
45 Percent	75	61-1/2 (Use 62)

Table 1-1 (Repeated)

Approximate Entering Conditions — Cooling Coil 10 Percent Outdoor Air, No Return-side Duct Gains				
Relative Humidity	Humid Climate		Dry Climate	
	Entering db	Entering wb	Entering db	Entering wb
55 Percent	77	65-1/2	77	64
50 Percent	77	64	77	62-1/2
45 Percent	77	62-1/2	77	61

Refer to **Manual P** for a procedure that can be used to determine precisely the condition of a mixture of return air and outdoor air.

Table 1-2 (Repeated)

3-8 Return-side Duct Gains

The condition of the air entering the indoor coil is affected by a return-side duct gain, which could be a sensible-only gain if a tightly sealed return duct is installed in an unconditioned space. It could also be a combination of sensible and latent gains if a leaky return duct is installed in an unconditioned space. (There are no gains associated with return ducts that are installed within the conditioned space.)

The information provided by Table 3-1 shows the effect that a return-side duct gain can have on the dry-bulb and wet-bulb temperature of the return air. The top part of this table shows

Return-side Duct Gains				
Conduction through Duct Wall				
Approximate Effect on the Condition of the Return Air	Sensible Return-side Gain (Percentage of the Sensible Load)			
	5%	10%	15%	20%
Return db	+1	+2	+3	+4
Return wb	+1/3	+2/3	+1	+1-1/3

1) For conduction loss calculations see **Manual D**, Section 12-2.
2) Dry-bulb rise = (Return-side Duct Gain) / (1.1 x Blower CFM)
3) Wet-bulb rise can be read from the Psychrometric Chart.

Return-side Leakage (10 Percent)				
Temperature of the Air Surrounding the Return Duct	Effect on Condition of Return Air			
	Humid Climate		Dry Climate	
	Entering db	Entering wb	Entering db	Entering wb
95 °F	+2	+1-1/2	+2	0
115 °F	+4	+2	+4	+1/2
135 °F	+6	+2-1/2	+6	+1

1) For leakage loss calculations see **Manual D**, Section 12-3.
2) Refer to **Manual P** for a procedure that can be used to determine the condition of the return-air-ambient-air mixture for any amount of return-side leakage.

Note

When duct runs are installed in an unconditioned space, they should be sealed tightly and well insulated. This effort will yield two important benefits:

- The equipment size will be reduced because the sensible and latent loads associated with the duct runs will be minimized.

- It will be easier to size the cooling equipment because the dry-bulb and wet-bulb temperatures of the air that enters the cooling coil will be within 1 °F of the temperatures associated with the air returned from the conditioned space.

Table 3-1

that, depending on the size of the conduction gain, the dry-bulb temperature will increase by 1 °F to 4 °F and the wet-bulb temperature will increase by about 1 °F. The bottom part of this table shows that, depending on the condition of the air that surrounds the return duct, a 10 percent return-side leakage rate will increase the dry-bulb temperature of the return air by 2 °F to 6 °F and the corresponding wet-bulb temperature increase will range from 0 °F to 2 °F. And, if the conduction gain and the leakage gain occur simultaneously (which is normally the case), the leakage effect will have to be added to the conduction effect.

3-9 Design Conditions

The size of the cooling equipment must be based on the same temperature and humidity conditions that were used to calculated the **Manual J** loads. This one-to-one correspondence is necessary because direct expansion cooling equipment is sensitive to variations in the temperature of the fluid (air or water) that enters the condenser and the condition (dry-bulb and wet-bulb temperature) of the air that is flowing toward the indoor coil.

An estimate of the cooling CFM also is useful (see page 1-3) because the cooling capacity tables published by equipment manufacturers are normally organized according to the air flow through the indoor coil (smallest to largest). This way, a quick search for a package can be made on the basis of airflow. However, this search may produce two or more packages that operate at the desired air flow, so the final selection must be based on the manufacturer's sensible and latent capacity data.

3-10 Step-by-Step Sizing Procedure

The following procedure provides a systematic method for sizing residential cooling equipment. This method uses the estimated cooling CFM as the initial selection parameter. The sensible and latent capacity requirements are then used to verify that the performance of the candidate unit (or candidate units) is acceptable. If the capacity check demonstrates that the candidate cooling package is not compatible with the design loads, the next larger (or smaller) unit will usually provide the desired performance.

Step 1 — Produce a Set of Design Parameters
Most of the information that is required for sizing residential cooling equipment is generated by the **Manual J** procedure. This basic collection of equipment sizing parameters includes the following items:

- Outdoor dry-bulb temperature
- Indoor dry-bulb temperature
- Indoor wet-bulb temperature
- Sensible load from Procedure D on **Form J-1**
- Latent load from Procedure D in **Form J-1**

If a ventilation load or a duct gain is associated with the return-side of the air distribution system, the condition of the air that enters the cooling coil will not be equal to the condition of the indoor air. When this is the case, Tables 1-2 and 3-1 can be used to evaluate the condition of the air that approaches the cooling coil. This exercise will produce two additional equipment sizing parameters:

- Entering dry-bulb temperature
- Entering wet-bulb temperature

The performance of water-to-air equipment will be affected by the temperature of the water that enters the condenser. Therefore, if water-source equipment is featured, one of the following items will have to be added to the list of equipment sizing parameters:

- Ground water temperature (open system)
- Maximum loop-water temperature (closed system)

Step 2 — Estimate the Cooling CFM
A considerable number of data tables are required to summarize the performance of the equipment packages that are associated with a particular product line. Since these tables normally are arranged according to the blower CFM, the search for candidate packages can be based on a blower CFM value that will be compatible with the local humidity conditions. The procedure for generating this information is summarized below:

A) Calculate the sensible heat ratio value.

$$SHR = \frac{Sensible\ Load}{Sensible\ Load + Latent\ Load}$$

B) Use Table 1-4 to find a TD value that will be compatible with the sensible and latent loads.

Sensible Heat Ratio Versus TD Value			
SHR	LAT	Room db	TD
Below 0.80	54	75	21
0.80 to 0.85	56	75	19
Above 0.85	58	75	17

Table 1-4 (Repeated)

C) Use the sensible heat equation to generate a cooling CFM value.

$$CFM = \frac{Sensible\ Load\ (from\ Procedure\ D)}{1.1\ X\ TD}$$

Step 3 — Search for Candidate Packages

Search the manufacturer's application data sheets for equipment packages that operate at a nominal CFM that is approximately equal to the calculated CFM value. If this search produces two or more candidates, select the packages that have an adequate amount of total capacity and discard any packages that are short on total capacity. (The total capacity value is equal to the sum of the sensible and latent capacity values that are listed in the set of design parameters.)

Step 4 — Evaluate Candidate Performance

Start with the smallest candidate package that has an adequate amount of total capacity. Compare the tabulated capacity data with the **Manual J** sensible and latent loads. (Remember to use the Step-1 design parameters.)

A candidate is acceptable if the sensible capacity of the equipment is equal to or greater than the calculated sensible load and if the latent capacity of the equipment is equal to or greater than the calculated latent load. Remember that oversizing restrictions apply. They are summarized in Sections 3-3 and 3-4.

• When the candidate package is considerably short of sensible or latent capacity, check the capacity of the next larger package (operating at the nominal fan speed).

• When the capacity of the candidate system exceeds the oversizing limitations, check the capacity of the next smaller system (operating at the nominal fan speed).

Sometimes the tabulated sensible capacity or latent capacity of a candidate package will be a little less than the required capacity. But it still may be possible to use the package if a shortage of one type of capacity is balanced by an excess of the other type of capacity.

• If the equipment is slightly short of sensible capacity, the package may be adequate if it has some excess latent capacity. Add half of the excess latent capacity to the sensible capacity and compare this adjusted value with the sensible load.

• If the equipment is slightly short of sensible capacity, an increase in fan speed will produce more sensible capacity and less latent capacity. Check to see if the equipment will satisfy the sensible and latent loads when more air is forced through the cooling coil. (The medium fan speed is preferred because, after the system is installed, a low air flow problem might be corrected by an increase in fan speed.)

• If the equipment is slightly short of latent capacity, a decrease in fan speed will generate more latent capacity and less sensible capacity. Check to see if the equipment will satisfy the sensible and latent loads when less air is forced through the cooling coil. (Medium fan speed is preferred because, after the system is installed, an excessive air flow problem might be corrected by a decrease in fan speed.)

3-11 Blower CFM for Duct Sizing

The estimated cooling CFM described in Step 2 is used only to search for candidate packages. The air flow design value that is used for the duct sizing calculations should match the blower CFM value that is associated with the equipment that will ultimately be installed. (Refer to ACCA **Manual D** for more information).

3-12 Interpolation

Interpolation of the manufacturer's cooling performance tables may be necessary when the actual design conditions do not closely match the published performance data. The following guidelines apply:

Condensing Temperature
Interpolation is recommended if the actual temperature of the condensing fluid (outdoor air temperature or water temperature) deviates from the published value by 5 °F or more. For example, if the performance data documents performance at 85 °F and 95 °F, an interpolation would be required when the outdoor air temperature is equal to 90 °F.

Dry-bulb at the Indoor Coil
Interpolation is recommended if the dry-bulb temperature of the entering air is more than 1 °F above or below the published value. It is not required if the dry-bulb temperature of the air approaching the indoor coil is within 1 °F of the published value.

Wet-bulb at the Indoor Coil
Interpolation is recommended if the wet-bulb temperature of the entering air is more than 1 °F above or below the published value. It is not required if the wet-bulb temperature of the air approaching the indoor coil is within 1 °F of the published value.

3-13 Example — Air-to-Air, No Return-side Gains

This example pertains to a system that features air-to-air equipment and duct runs installed within the conditioned space. In this case the entire list of equipment sizing parameters can be extracted from the **Manual J** load calculation data (see Step 1 below).

Step 1 — Design Parameters

• Outdoor dry-bulb = 95 °F
• Indoor dry-bulb = 75 °F
• Indoor relative humidity = 50 percent
• Indoor wet-bulb = 63 °F

Model 25		Outdoor Air Temperature				
		95 °F				
Enter Wet-Bulb (°F)	Total Air Flow (CFM)	Total Cool Cap. (BTUH)	Comp. Motor Watts Input	Sensible-to-Total Ratio (S/T)		
				Dry Bulb °F		
				76	80	84
63 °F	1,050	34,400	3,510	.76	.89	.99
	1,200	35,400	3,570	.80	.93	1.00
	1,350	36,400	3,610	.84	.97	1.00
CFM = 1,200 Total Capacity = 35,400 BTUH Total Load = 33,900 BTUH Percent Oversize = 35,400 / 33,900 = 1.04 (acceptable)						

Table 3-2

Model 30		Outdoor Air Temperature				
		95 °F				
Enter Wet-Bulb (°F)	Total Air Flow (CFM)	Total Cool Cap. (BTUH)	Comp. Motor Watts Input	Sensible-to-Total Ratio (S/T)		
				Dry Bulb °F		
				76	80	84
63 °F	1,200	39,700	3,870	.75	.88	.98
	1,325	40,700	3,910	.78	.91	1.00
	1,575	41,500	3,940	.80	.94	1.00
CFM = 1,325 Total Capacity = 40,700 BTUH Total Load = 33,900 BTUH Percent Oversize = 40,700 / 33,900 = 1.20 (acceptable)						

Table 3-3

- Sensible load in conditioned space = 27,200 BTUH
- Latent load in conditioned space = 6,700 BTUH
- Total load = 33,900 BTUH
- Outdoor air CFM (ventilation) = none

Step 2 — Approximate Cooling CFM

$$SHR = \frac{27200}{27200 + 6700} = 0.80$$

$$TD = 19 \,°F$$

$$CFM = \frac{27200}{1.1 \times 19} = 1300$$

Step 3 — A performance data survey based on the 1,300 CFM air flow value and 33,900 BTUH of total capacity produced two candidate packages. Table 3-2 shows that when the Model 25 package operates at 1,200 CFM, it will have about 4 percent excess capacity. And Table 3-3 shows that when the Model 30 package operates at 1,325 CFM, it will have about 20 percent excess capacity.

Step 4 — Table 3-4 indicates that both packages have an adequate amount of sensible and latent capacity. This summary also indicates that both packages comply with the oversizing rule. Therefore, either package is acceptable because they both satisfy all of the sizing guidelines. (In this example, the 76 °F dry-bulb temperature in the performance data does not exactly match the 75 °F db design condition, but this difference will not have a significant effect on the capacity of the equipment.)

3-14 Example — Air-to-Air, Excess Latent Capacity

This example is similar to the previous one, except that the sensible load is a little larger and the latent load is noticeably

Performance Summary						
	CFM	TC	S/T	SC	LC	Over
Model 25	1,200	35,400	0.80	28,320	7,080	1.04
Model 30	1,325	40,700	0.78	31,746	8,954	1.20
Sensible Load = 27,200 BTUH Latent Load = 6,700 BTUH						

Table 3-4

smaller. The values for the sensible and latent loads and the other sizing parameters are summarized in Step 1 below:

Step 1 — Design Parameters

- Outdoor design dry-bulb = 95 °F
- Indoor dry-bulb = 75 °F
- Indoor relative humidity = 50 percent
- Indoor wet-bulb = 63 °F
- Sensible load in conditioned space = 30,100 BTUH
- Latent load in conditioned space = 3,100 BTUH
- Total Load = 33,200 BTUH
- Outdoor air CFM (ventilation) = none

Step 2 — Approximate Cooling CFM

$$SHR = \frac{30100}{30100 + 3100} = 0.91$$

$$TD = 17 \,°F$$

$$CFM = \frac{30100}{1.1 \times 17} = 1600$$

Step 3 — In this case, a performance data survey based on the 1,600 CFM air flow value suggested the Model 50 equipment

Model 50	Outdoor Air Temperature					
	95 °F					
Enter Wet-Bulb (°F)	Total Air Flow (CFM)	Total Cool Cap. (BTUH)	Comp. Motor Watts Input	Sensible to Total Ratio (S/T)		
				Dry Bulb °F		
				76	80	84
63 °F	1,400	45,400	4,290	.77	.90	1.00
	1,600	46,800	4,360	.81	.95	1.00
	1,800	48,100	4,420	.85	.98	1.00

CFM = 1,600
CFM / TON = (1,600 x 12,000) / 46,800 = 410
Total Capacity = 46,800 BTUH
Total Load = 33,200 BTUH
Percent Oversize = 46,800 / 33,200 = 1.41 (too big)

Table 3-5

AC-36	Outdoor Air Temperature			
	95 °F			
Enter Wet-Bulb (°F)	Total Air Flow (CFM)	Total Cool Cap. (BTUH)	Sensible Capacity (BTUH)		
			Dry Bulb °F		
			76	78	80
59 °F		31,800	30,700	31,400	31,800
63 °F	1600	34,200	31,000	33,200	34,200
67 °F		36,700	25,300	28,400	31,600

CFM = 1,600
CFM / Ton = (1,600 x 12,000) / 34,200 = 561
Total Capacity = 34,200 BTUH
Total Load = 33,200 BTUH
Percent Oversize = 34,200 / 33,200 = 1.03 (acceptable)

Table 3-6

package, but this equipment has too much total capacity, as indicated by Table 3-5. However, the table also indicates that the equipment is operating at about 400 CFM of air flow per ton of total capacity. In other words, this equipment package features a condensing-unit-blower-coil combination that is compatible with an application that needs a relatively cold evaporator coil (for dehumidifying capacity).

For this particular set of design loads, the equipment package should operate at about 530 CFM of air flow per ton of total capacity. Unfortunately, some manufacturers publish data only for condensing-unit-blower-coil combinations that are setup to operate between 400 and 450 CFM per ton. In these cases, the total capacity requirement should be used to search for a candidate package. Therefore, the performance of the Model 25 package should be evaluated because it has the correct amount of total capacity (see Table 3-2).

Note that some manufacturers do publish performance data for a wide range of condensing-unit-blower-coil combinations. In some cases, the tabulated air-flow-to-capacity ratios range from less than 400 CFM per ton to more than 600 CFM per ton. An example of this type of data is provided by Table 3-6, which indicates that the AC-36 package has the desired blower CFM value and the correct amount of total capacity. (In this case, the manufacturer provides documentation on a condensing-unit-blower-coil combination that operates at about 560 CFM per ton of total capacity)

Step 4 — If the Model 25 package is installed in this home, about one half of the excess latent capacity will be used to remove moisture from the indoor air and the other half will be converted to sensible capacity. (Since there is excess latent capacity, the indoor humidity will be less than the 50 percent design value and the entering wet-bulb temperature will be lower than 63 °F.) The following calculations indicate that the Model 25 package has an adequate amount of sensible and latent capacity when it operates at 1,200 CFM:

Sensible Capacity = 28320 BTUH
Latent Capacity = 7080 BTUH
Latent Load = 3100 BTUH
Excess Latent Capacity = 7080 - 3100 = 3980 BTUH
Additional Sensible Capacity = 0.50 x 3980 = 1990 BTUH
Sensible Capacity = 28320 + 1990 = 30310 BTUH (acceptable)
Latent Capacity = 35400 - 30310 = 5090 (RH less than 50%)

The AC-36 package also is a viable candidate. Table 3-7 compares the performance of the AC-36 package with the performance of the Model 25 package. This summary shows that either equipment package is acceptable because they both satisfy all the sizing criteria. (At 75 °F dry-bulb, the comfort zone ranges from less than 40 percent RH to more than 55 percent RH.)

Performance Summary					
	CFM	TC	SC	LC	Indoor RH
Model 25	1,200	35,400	30,310	5,090	Below 50%
AC-36	1,600	34,200	31,000	3,200	50%
Sensible Load = 30,100 BTUH Latent Load = 3,100 BTUH					

Table 3-7

Also note that the performance of the Model 25 package is compatible with the cooling loads when the blower is operated at high speed. This is demonstrated on the next page by Table 3-8, which compares the high-speed performance with the sensible and latent loads. This table shows that the unit has more than enough sensible capacity and that the indoor humidity will be lower than the design value. However, the ability to use a blower speed adjustment to compensate for a restriction in the air distribution system will be sacrificed.

Model 25 — High-speed Blower					
	CFM	TC	SC	LC	Indoor RH
Model 25	1,350	36,400	30,576	5,824	Below 50%
Loads	NA	33,200	30,100	3,100	50%

Table 3-8

3-15 Example — Air-to-Air, Mechanical Ventilation

This example is similar to example 3-13, except it shows how to use a manufacturer's performance data to size cooling equipment when outdoor air is introduced through the return-side of the duct system. The values for the sensible and latent loads and the other sizing parameters associated with this example are summarized in Step 1 below. Also note that all the duct runs will be installed within the conditioned space. (The procedure presented in this example does not apply if the outdoor air CFM exceeds 10 percent of the blower CFM. Refer to **Manual P** for a procedure that can be used to evaluate the condition of the entering air if the outdoor air flow exceeds this 10 percent limit.)

Step 1 — Design Parameters

- Outdoor dry-bulb = 95 °F (humid climate)
- Indoor dry-bulb = 75 °F
- Indoor relative humidity = 50 percent
- Indoor wet-bulb = 63 °F
- Sensible load in conditioned space = 27,200 BTUH
- Latent load in conditioned space = 6,700 BTUH
- Ventilation rate = 100 CFM
- Sensible outdoor air load = 2,200 BTUH (Procedure D)
- Latent outdoor air load = 1,900 BTUH (Procedure D)
- Sensible design load = 29,400 BTUH (Procedure D)
- Latent design load = 8,600 BTUH (Procedure D)
- Total design load = 38,000 BTUH
- Entering dry-bulb = 77 °F (Table 1-2)
- Entering wet-bulb = 64 °F (Table 1-2)

Step 2 — Approximate Cooling CFM

$$SHR = \frac{29400}{29400 + 8600} = 0.77$$

$$TD = 21 \, ^{\circ}F$$

$$CFM = \frac{29400}{1.1 \times 21} = 1270 \text{ (rounded)}$$

Step 3 — A survey of the performance data, based on the 1,300 CFM air flow value and 38,000 BTUH of total capacity, disqualifies the Model 25 package because it does not have enough total capacity (see Table 3-2). Also note that the 1,600 CFM, AC-36 package also is short on total capacity (see Table 3-6).

However, Table 3-9 shows that when the Model 30 package operates at 1,325 CFM, it qualifies as a candidate package because it has a few thousand extra BTUH of total capacity. (In this case, Table 1-2 applies because 100 CFM of outdoor air is almost 10 percent of the blower CFM.)

Model 30		Outdoor Air Temperature				
		95 °F				
Enter Wet-Bulb (°F)	Total Air Flow (CFM)	Total Cool Cap. (BTUH)	Comp. Motor Watts Input	Sensible to Total Ratio (S/T)		
				Dry Bulb °F		
				76	77	80
63 °F	1,200	39,700		.75		.88
	1,325	40,700		.78		.91
	1,575	41,500		.80		.94
64 °F	1,325	41,300		.735	.766	.858
67 °F	1,200	42,200		.59		.68
	1,325	43,100		.60		.70
	1,575	43,900		.62		.74
CFM = 1,325 Percent Outdoor Air = 100 / 1,325 = 7-1/2 (Use 10%) Total Capacity (@ 77/64) = 41,300 BTUH Total Load = 38,000 BTUH Percent Oversize = 41,300 / 38,000 = 1.07 (acceptable)						

Table 3-9

Step 4 — Table 3-10 indicates that the Model 30 package has an adequate amount of sensible and latent capacity for this application. In fact, the equipment has some excess latent capacity, so the indoor humidity will be a little lower than the 50 percent design value.

Interpolated Performance Summary — Model 30					
Ventilation	CFM	TC	SC	LC	Indoor RH
100 CFM	1,325	41,300	31,600	9,700	Below 50%
Sensible Load = 29,400 BTUH Latent Load = 8,600 BTUH					

Table 3-10

3-16 Example — Air-to-Air, Duct Gains

This example shows that duct runs must be sealed and generously insulated when they are installed in an unconditioned space. Otherwise, duct gains will have a significant effect on comfort, the design loads, the performance of the cooling equipment, and the cooling capacity requirements.

Except for the location of the duct system (attic), this example is the same as the first example (Section 3-13). The values for the sensible and latent loads, duct gains, and the other sizing parameters associated with this example are summarized in Step 1 below. Note that outdoor air is not used for mechanical ventilation, but there is an outdoor air load created by the return-side duct leakage. (See **Manual D** for information about duct gain calculations.)

Step 1 — Design Parameters

- Outdoor air = 95 °F, 99 Grains (humid climate)
- Grains difference = 34 (**Manual J**, Table 1)
- Indoor air = 75 °F, 62-1/2 WB (65 Grains)
- Indoor relative humidity = 50 percent
- Air off coil = 55 db, 56 Grains (estimate)
- Sensible load in conditioned space = 27,200 BTUH
- Latent load in conditioned space = 6,700 BTUH
- Ventilation rate = no provision (0 CFM)
- Duct installed in attic, 2-inch blanket
- Air in attic = 125 °F db (field measurement)
- Estimated supply-side conduction gain = 1,000 BTUH
- Estimated supply-side leakage = 80 CFM
- Estimated return-side conduction gain = 1,200 BTUH
- Estimated return-side leakage = 100 CFM

Step 1A — Duct Leakage Load (Rounded)

Sensible load = 1.1 x 80 x (125 - 55) = 6160 BTUH
Latent load = 0.68 x 80 x (99 - 56) = 2340 BTUH

Step 1B — Return-side Ventilation Effect (Rounded)

Sensible load = 1.1 x (100 - 80) x (125 - 75) = 1100 BTUH
Latent load = 0.68 x (100 - 80) x (99 - 65) = 460 BTUH

Step 1C — Equipment Sizing Loads (BTUH)

Sensible = 27200 + 1000 + 1200 + 6160 + 1100 = 36660
Latent load = 6700 + 2340 + 460 = 9500
Total load = 36660 + 9500 = 46160

Step 2 — Approximate Cooling CFM

$$SHR = \frac{36660}{46160} = 0.79$$

$$TD = 21 \text{ °F}$$

$$CFM = \frac{36660}{1.1 \times 21} = 1590$$

Step 3 — In this case, a survey based on the 1,600 CFM air flow value and the 46,160 BTUH total load value leads to the Model 50 package. Table 3-11 shows that when the Model 50 package operates at 1,600 CFM, it qualifies as a candidate because it has an ample amount of total capacity.

Since this design features a system that is penalized by return-side duct gains, the condition of the entering air must be evaluated. Table 3-1 provides the information that is required for this work. These calculations are summarized below.

Conduction loss percentage = 1200 / 36660 = 3% (use 5%)
Conduction loss adjustment — dry-bulb = +1 °F
Conduction loss adjustment — wet-bulb = +1/3 °F
Percent return-side leakage = 100 / 1600 = 6%
Leakage adjustment @ 125 °F = 0.6 x 5 =+3 °F db
Leakage adjustment @ 125 °F = 0.6 x 2.25 =+1-1/3 °F wb
Total dry-bulb adjustment = 1 + 3 = 4 °F
Total wet-bulb adjustment = 1/3 + 1-1/3 = 2 °F (rounded)
Entering dry-bulb = 75 + 4 = 79 °F
Entering wet-bulb = 63 + 2 = 65 °F

Model 50		Outdoor Air Temperature					
		95 °F					
Enter Wet-Bulb (°F)	Total Air Flow (CFM)	Total Cool Cap. (BTUH)	Comp. Motor Watts Input	Sensible to Total Ratio (S/T)			
				Dry Bulb °F			
				76	79	80	
63 °F	1,400	45,400		.77		.90	
	1,600	46,800		.81		.95	
	1,800	48,100		.85		.98	
65 °F	1,600	48,200		.72	.818	.85	
67 °F	1,400	48,200		.61		.72	
	1,600	49,600		.63		.75	
	1,800	50,700		.65		.78	
CFM = 1,600							
Total Capacity (@ 79/65) = 48,200 BTUH							
Total Load = 46,160 BTUH							
Percent Oversize = 48,200 / 46,160 = 1.04 (acceptable)							

Table 3-11

Step 4 — On the next page, Table 3-12 indicates that the Model-50 package has enough sensible capacity and that it is about 730 BTUH short on latent capacity. This table also shows the Model 25 performance summary. This package, which is much smaller, would be acceptable if the duct runs were located within the conditioned space. (If the ducts must be installed in the attic, all the seams should be sealed with durable materials and the R-value of the insulation should comply with the guidelines published in the Council of Building Officials Model Energy Code (CABO-MEC) or the American Society of Heating, Refrigeration and Air Conditioning Engineers (ASHRAE) Standard 90.2.)

Performance Summary — Model 50					
Duct Gains	CFM	TC	SC	LC	Indoor RH
Yes	1,600	48,200	39,430	8,770	About 50%
Sensible Load = 36,660 BTUH Latent Load = 9,500 BTUH					

Performance Summary — Model 25					
Duct Gains	CFM	TC	SC	LC	Indoor RH
No	1,200	35,400	28,320	7,080	About 50%
Sensible Load = 27,200 BTUH Latent Load = 6,700 BTUH					

Table 3-12

It is possible to create an example problem that combines a ventilation load with a return-side duct leakage load. However, this problem would be similar to the previous example if the total flow of outdoor air (ventilation plus return-side leakage) is less than 10 percent of the blower CFM.

3-17 Example — Open Water-to-Air System

The procedure used to select water-cooled equipment is almost identical to the procedure used to select air-cooled equipment. The only difference is that the equipment performance is not affected by the outdoor air temperature. In this case, the cooling capacity tables are based on the temperature of the water that flows through the condenser.

When a open system is installed, ground water (usually well water) makes one pass through the condenser. This means that the entering water temperature will be equal to the temperature of the ground water. This temperature will range between 40 °F and 70 °F, depending upon the local climate. In any case, the performance of the once-through water cooled system will be superior to the performance of an air cooled system because the temperature of the cooling fluid is so much lower.

To create a basis for comparison, the list of equipment sizing parameters for this water-to-air example will be identical to the list associated with the first air-to-air example (Section 3-13). This information is summarized in Step 1 below.

Step 1 — Design Parameters

- Outdoor dry-bulb = 95 °F
- Temperature of ground water = 50 °F
- Indoor dry-bulb = 75 °F

- Indoor relative humidity = 50 percent
- Indoor wet-bulb = 63 °F
- Sensible load in conditioned space= 27,200 BTUH
- Latent load in conditioned space= 6,700 BTUH
- Total load = 33,900 BTUH
- Outdoor air CFM (ventilation) = none

Step 2 — Approximate Cooling CFM

$$SHR = \frac{27200}{27200 + 6700} = 0.80$$

$$TD = 19\ ^{\circ}F$$

$$CFM = \frac{27200}{1.1 \times 19} = 1300$$

Step 3 — A performance data survey based on the 1,300 CFM air flow value and a minimum water-side flow rate value (GPM) produced two possible candidates. Table 3-13 shows that the W-36 package is a candidate because when it operates at 1,200 CFM, it has about 14 percent excess capacity. But Table 3-14 shows that the W-41 package is not a candidate because it has too much cooling capacity. (When a once-through system is designed, the costs associated with installing and operating the water side of the system can be minimized if the equipment selection is based on the lowest water flow rate value.)

W-36		Cooling Capacity Data			
		1,200 CFM			
EWT	GPM	Entering db / wb	Total Capacity	Sensible Capacity	Input Kw
50	4.5	75/63	38.7	27.6	2.67
		80/67	42.1	28.7	2.72
		85/71	45.4	29.9	2.80
Total Capacity = 38,700 BTUH Total Load = 33,900 BTUH Percent Oversize = 38,700 / 33,900 = 1.14 (acceptable)					

Table 3-13

W-41		Cooling Capacity Data			
		1,375 CFM			
EWT	GPM	Entering db / wb	Total Capacity	Sensible Capacity	Input Kw
50	5.0	75/63	44.4	31.1	3.32
		80/67	48.2	32.4	3.39
		85/71	52.1	33.7	3.45
Total Capacity = 44,400 BTUH Total Load = 33,900 BTUH Percent Oversize = 44,400 / 33,900 = 1.31 (not acceptable)					

Table 3-14

Step 4 — Table 3-15 indicates that the W-36 package has an adequate amount of sensible capacity and more than 4,000 BTUH of extra latent capacity, which means that the indoor humidity will be a little lower than the 50 percent design value. Therefore, this package is acceptable because it satisfies all of the sizing guidelines.

Performance Summary					
W-36	**CFM**	**TC**	**SC**	**LC**	**Over**
	1,200	38,700	27,600	11,100	1.14
Sensible Load = 27,200 BTUH Latent Load = 6,700 BTUH					

Table 3-15

3-18 Example — Buried Water Loop System

When the water side of an water-to-air system features a buried piping loop, the temperature of the water in the loop could exceed 90 °F by the end of the cooling season. Therefore, when the cooling package is selected, the rated capacity must be correlated with this late summer water temperature. Otherwise, the procedure used to select equipment for a water-loop system is identical to the procedure used to select equipment for a once-through system.

An approximate value for the maximum water temperature in a closed loop system can be estimated by subtracting 10 °F from the hottest summer temperatures that will occasionally occur at the home site. (For most locales, this temperature is about 5 °F to 15 °F hotter than the **Manual J** design temperature.)

Also note that the minimum flow rate through the water side of the equipment package is dictated by the water velocity that is associated with the piping loop. (The flow in the piping loop must be turbulent.) This value is normally generated by the computer program (or hand calculations) that is used to design the piping system.

For this example, the list of equipment sizing parameters will be identical to the list that was used for the previous example, except that values for the maximum summer temperature and the minimum water-side flow rate have been added to the data set. This information is summarized in Step 1 below.

Step 1 — Design Parameters

- **Manual J** design temperature = 95 °F
- Maximum summer temperature = 100 °F
- Maximum water temperature = 90 °F
- Minimum water-side flow rate = 7 GPM
- Indoor dry-bulb = 75 °F
- Indoor relative humidity = 50 percent

- Indoor wet-bulb = 63 °F
- Sensible load in conditioned space = 27,200 BTUH
- Latent load in conditioned space = 6,700 BTUH
- Total load = 33,900 BTUH
- Outdoor air CFM (ventilation) = none

Step 2 — Approximate Cooling CFM

$$SHR = \frac{27200}{27200 + 6700} = 0.80$$

$$TD = 19\ ^oF$$

$$CFM = \frac{27200}{1.1 \times 19} = 1300$$

Step 3 — A performance data survey based on the 1,300 CFM air flow value and a 7 GPM water-side flow rate value will lead to the same equipment packages evaluated in the previous example. But in this case, Table 3-16 shows that the W-36 package is not a candidate because, when it operates at 1,200 CFM, it does not have enough total capacity. And Table 3-17 (see next page) shows that the W-41 package is a candidate because it has the right amount of total capacity.

This reversal of circumstances (the W-36 model was the package that was compatible with the once-through water system) is related to the temperature of the condenser water. In this case, the entering water temperature is 40 °F warmer than the temperature used in the previous example. This difference translates into a sizable reduction in cooling capacity. (Compare the Table 3-13 data with the Table 3-16 data.)

W-36		Cooling Capacity Data			
		1,200 CFM			
EWT	GPM	Entering db / wb	Total Capacity	Sensible Capacity	Input Kw
90	4.5	75/63	29.9	22.7	3.39
		80/67	32.5	23.6	3.46
		85/71	35.1	24.6	3.53
	7.0	75/63	30.8	23.1	3.29
		80/67	33.5	24.1	3.36
		85/71	36.1	25.1	3.42
	9.0	75/63	31.7	23.6	3.19
		80/67	34.5	24.6	3.26
		85/71	37.3	25.6	3.33
Total Capacity = 30,800 BTUH Total Load = 33,900 BTUH Percent Oversize = 30,800 / 33,900 = 0.91 (too small)					

Table 3-16

W-41		Cooling Capacity Data			
		1,375 CFM			
EWT	GPM	Entering db / wb	Total Capacity	Sensible Capacity	Input Kw
90	5.0	75/63	35.6	27.3	4.29
		80/67	38.7	28.4	4.38
		85/71	41.8	29.6	4.47
	8.0	75/63	36.3	27.6	4.09
		80/67	39.5	28.7	4.17
		85/71	42.7	29.9	4.25
	11	75/63	37.1	27.8	3.97
		80/67	40.3	29.0	4.05
		85/71	43.5	30.2	4.13

Total Capacity = 36,300 BTUH
Total Load = 33,900 BTUH
Percent Oversize = 36,300 / 33,900 = 1.07 (acceptable)

Table 3-17

Step 4 — Table 3-18 indicates that the W-41 package has just enough sensible capacity and about 2,000 BTUH of extra latent capacity. And, the unit is only oversized by 7 percent when the system is subjected to the design conditions. Therefore, this package is acceptable because it satisfies all of the sizing guidelines.

Performance Summary					
W-41	CFM	TC	SC	LC	Over
	1,375	36,300	27,600	8,700	1.07

Sensible Load = 27,200 BTUH
Latent Load = 6,700 BTUH

Table 3-18

Bear in mind that the sizing guidelines that pertain to water-loop systems are based on a worst-time-of-year scenario. This means that the equipment will have additional capacity during most of the cooling season because the entering water temperature will not approach the 90 °F design value until the very end of the summer. If fact, in May or June it will be lower than the 50 °F temperature that was used to design the once-through system.

3-19 Water-to-Air, Excess Latent Capacity

As far as the air-side performance is concerned, water-to-air equipment is no different than air-to-air equipment. This means that when there is too much latent capacity, the indoor humidity and the entering air temperature will be lower than the design value. This, in turn, will cause a slight reduction in total capacity and an increase in sensible capacity. (About half of the extra latent capacity will be converted to sensible capacity. Refer to Sections 3-6 and 3-14 for a review of these concepts.)

3-20 Water-to-Air, Mechanical Ventilation

This example is similar to the Section 3-17 example (open water system), except it shows how to use a manufacturer's performance data to size water-to-air equipment when outdoor air is introduced through the return-side of the duct system. The values for the sensible and latent loads and the other sizing parameters associated with this example are summarized below. Also note that all the duct runs will be installed within the conditioned space. (The procedure presented in this example does not apply if the outdoor air CFM exceeds 10 percent of the blower CFM. Refer to **Manual P** for a procedure that can be used to evaluate the condition of the entering air if the outdoor air flow exceeds this 10 percent value.)

Step 1 — Design Parameters

- Outdoor dry-bulb = 95 °F (humid climate)
- Temperature of ground water = 50 °F
- Indoor dry-bulb = 75 °F
- Indoor relative humidity = 50 percent
- Indoor wet-bulb = 63 °F
- Sensible load in conditioned space = 27,200 BTUH
- Latent load in conditioned space = 6,700 BTUH
- Ventilation rate = 100 CFM
- Sensible outdoor air load = 2,200 BTUH (Procedure D)
- Latent outdoor air load = 1,900 BTUH (Procedure D)
- Sensible design load = 29,400 BTUH (Procedure D)
- Latent design load = 8,600 BTUH (Procedure D)
- Total design load = 38,000 BTUH

Step 2 — Approximate Cooling CFM

$$SHR = \frac{29400}{29400 + 8600} = 0.77$$

$$TD = 21 \ ^\circ F$$

$$CFM = \frac{29400}{1.1 \times 21} = 1270 \ (rounded)$$

Step 3 — A survey of the performance data— based on the 38,000 BTUH load, 1,270 CFM of air flow, and a minimum water-side flow rate value — leads to the W-36 package because it seems to have just enough total capacity (see Table 3-13). However, an interpolated capacity check is required because the cooling performance must be evaluated at the 77 °F dry-bulb, 64 °F wet-bulb condition that is suggested by Table 1-2. (Note that 100 CFM of outdoor air is almost 10 percent of the blower CFM.) In this regard, Table 3-19 shows that the W-36 will have a little more total capacity when it is subjected to these entering conditions. (The interpolation is based on wb because of the format of the manufacturer's data table.)

| W-36 | | Cooling Capacity Data | | | |
| | | 1,200 CFM | | | |
EWT	GPM	Entering db / wb	Total Capacity	Sensible Capacity	Input Kw
50	4.5	75/63	38.7	27.6	2.67
		77/64	40.4	28.1	
		80/67	42.1	28.7	2.72
Approximate Total Capacity = 40,400 BTUH Total Load = 38,000 BTUH Percent Oversize = 40,400 / 38,000 = 1.06 (acceptable)					

Table 3-19

Step 4 — Table 3-20 indicates that the W-36 package is about 1,300 BTUH short on sensible capacity and that it has about 3,700 BTUH of extra latent capacity. Therefore, the equipment will have enough sensible capacity if half of this extra latent capacity is converted to sensible capacity.

Performance Summary — W-36 Model					
Ventilation	CFM	TC	SC	LC	Indoor RH
100 CFM	1,200	40,400	28,100	12,300	Below 50%
Sensible Load = 29,400 BTUH Latent Load = 8,600 BTUH					

Table 3-20

If this example had featured a buried-loop piping system, the entering water temperature would be much warmer than the 50 °F value associated with the open system. This means that the W-36 package, which was just adequate with 4.5 GPM and 50 °F water, would not have enough cooling capacity if the water temperature was in the 90 °F range. Therefore, a larger equipment package would be required — the W-41 package operating at 11 GPM, for example (see Table 3-17).

3-21 Example — Water-to-Air, Duct Gains

It has already been demonstrated that return-side duct gains will have a significant effect on the design loads and the performance of air-to-air cooling equipment (refer to Section 3-16). This example shows that return-side duct gains have the same effect on a water-to-air system. The values for the sensible and latent loads and the other sizing parameters associated with this example are summarized below. Except for the information that pertains to the duct gains, this example is similar to example 3-17, which features an open water system. (See **Manual D** for information about duct gain calculations.)

Step 1 — Design Parameters

- Outdoor dry-bulb = 95 °F, 99 Grains (humid climate)
- Grains difference = 34 (**Manual J**, Table 1)
- Temperature of the ground water = 50 °F
- Indoor dry-bulb = 75 °F, 62-1/2 WB (65 Grains)
- Indoor relative humidity = 50 percent
- Air off coil = 55 db, 56 Grains (estimate)
- Sensible load in conditioned space = 27,200 BTUH
- Latent load in conditioned space = 6,700 BTUH
- Ventilation rate = no provision (0 CFM)
- Duct installed in attic, 2 inch blanket
- Air in attic = 125 °F db (field measurement)
- Estimated supply-side conduction gain = 1,000 BTUH
- Estimated supply-side leakage = 80 CFM
- Estimated return-side conduction gain = 1,200 BTUH
- Estimated return-side leakage = 100 CFM

Step 1A — Duct Leakage Load (Rounded)

Sensible load = 1.1 x 80 x (125 - 55) = 6160 BTUH
Latent load = 0.68 x 80 x (99 - 56) = 2340 BTUH

Step 1B — Return-side Ventilation Effect (Rounded)

Sensible load = 1.1 x (100 - 80) x (125 - 75) = 1100 BTUH
Latent load = 0.68 x (100 - 80) x (00 - 65) = 460 BTUH

Step 1C — Equipment Sizing Loads (BTUH)

Sensible = 27200 + 1000 + 1200 + 6160 + 1100 = 36660
Latent load = 6700 + 2340 + 460 = 9500
Total load = 36660 + 9500 = 46160 BTUH

Step 2 — Approximate Cooling CFM

$$SHR = \frac{36660}{46160} = 0.79$$

$$TD = 21 \; °F$$

$$CFM = \frac{36660}{1.1 \times 21} = 1590$$

Step 3 — A cursory survey of the performance data — based on a 1,600 CFM air flow value, a 46,160 BTUH load, and a minimum water-side flow rate — leads to a W-49 package. The performance of this equipment package is summarized by Table 3-21. But in this case the system is penalized by return-side duct gains, so an entering dry-bulb and wet-bulb temperature adjustment is required. Table 3-1 provides the information required for this work. These calculations are summarized below.

Conduction loss percentage = 1200 / 36660 = 3% (use 5%)
Conduction loss adjustment — dry-bulb = +1 $^{\circ}$F
Conduction loss adjustment — wet-bulb = +1/3 $^{\circ}$F
Percent return-side leakage = 100 / 1600 = 6%
Leakage adjustment @ 125 $^{\circ}$F = 0.6 x 5 = +3 $^{\circ}$F db
Leakage adjustment @ 125 $^{\circ}$F = 0.6 x 2.25 = +1-1/3 $^{\circ}$F wb
Total dry-bulb adjustment = 1 + 3 = 4 $^{\circ}$F
Total wet-bulb adjustment = 1/3 + 1-1/3 = 2 $^{\circ}$F (rounded)
Entering dry-bulb = 75 + 4 = 79 $^{\circ}$F
Entering wet-bulb = 63 + 2 = 65 $^{\circ}$F

In this case, an interpolated capacity check is required because the cooling performance must be evaluated at the 79 $^{\circ}$F dry-bulb, 65 $^{\circ}$F wet-bulb condition. Table 3-21 indicates that the W-49 package will have an adequate amount of total capacity. (The interpolation is based on wb because of the format of the manufacturer's data table.)

W-49		Cooling Capacity Data			
		1,600 CFM			
EWT	GPM	Entering db / wb	Total Capacity	Sensible Capacity	Input Kw
50	6.0	75/63	51.0	35.9	3.97
		79/65	53.2	36.7	
		80/67	55.4	37.4	4.05
	9.0	75/63	52.6	36.6	3.86
		80/67	57.1	38.1	3.93
90	6.0	75/63	41.3	31.3	4.98
		80/67	44.9	32.6	5.08
	9.0	75/63	42.6	31.9	4.84
		80/67	46.3	33.2	4.93
	12.0	75/63	43.9	32.5	4.69
		80/67	47.7	33.9	4.79

Total Capacity = 53,200 BTUH
Total Load = 46,160 BTUH
Percent Oversize = 53,200 / 46,160 = 1.15 (acceptable)

Table 3-21

Step 4 — Table 3-22 indicates that the W-49 package has enough sensible capacity and that it is about 7,000 BTUH of extra latent capacity. Therefore the indoor relative humidity will be lower than 50 percent value and about 3,500 BTUH of latent capacity will be converted to sensible capacity. This figure also shows the W-36 performance summary. This package, which is much smaller, would be acceptable if the duct runs were located within the conditioned space. (If the ducts must be installed in the attic, all the seams should be sealed with durable materials and the R-value of the insulation should comply with the guidelines that are published in the Council of Building Officials Model Energy Code (CABO-MEC) or the American Society of Heating, Refrigeration and Air Conditioning Engineers (ASHRAE) Standard 90.2.)

Performance Summary — W-49 Package					
Duct Gains	CFM	TC	SC	LC	Indoor RH
Yes	1,600	53,200	36,700	16,500	Below 50%

Sensible Load = 36,660 BTUH
Latent Load = 9,500 BTUH

Performance Summary — W-36 Package					
Duct Gains	CFM	TC	SC	LC	Indoor RH
No	1,200	38,700	27,600	11,100	Below 50%

Sensible Load = 27,200 BTUH
Latent Load = 6,700 BTUH

Table 3-22

If this example had featured a buried-loop piping system, the entering water temperature would be much warmer than the 50 $^{\circ}$F value associated with the open system. This means that the W-49 package, which was just adequate with 6 GPM and 50 $^{\circ}$F water, would be short on total capacity and sensible capacity if the entering water temperature was in the 90 $^{\circ}$F range. In fact, the W-49 package would still be a few thousand BTUH short on total capacity and sensible capacity if the flow rate is increased to 12 GPM.

Section 4
Air-source Heat Pump

4-1 Equipment Size Is Based on the Cooling Load

In a cold climate, the heating load is usually larger than the cooling load, and the heating costs are usually much higher than the cooling costs. However, the cooling season cannot be ignored when the equipment is selected. So for any type of climate, the size of the heat pump should be based on the cooling load. Then after the equipment package is selected, a balance-point diagram can be used to evaluate the heating performance.

Some manufacturers do not publish detailed equipment performance data, but it may be available upon request. If the appropriate capacity information is not available, the designer is advised to use a product supported by the necessary documentation. (Refer to Section 8 of this manual for an explanation of why the ARI rating data should not be used for equipment selection.)

4-2 Manufacturer's Application Data

The equipment selection procedure requires comprehensive rating tables that specify the cooling performance of specific combinations of outdoor packages and indoor fan coil units. This performance data is similar to the "cooling only" tables that appear in Sections 3-13, -14, -15 and -16 of this manual.

Tables that document the heating performance over a full range of outdoor temperatures also are required. Because air-to-air equipment is subject to outdoor coil icing, these tables should include a defrost cycle adjustment. (Sometimes the term "integrated capacity" is used to describe the net amount of heating capacity that will be available after an allowance is made for the cooling effect that occurs during the defrost cycle.) Table 4-1 provides an example of tabulated heating performance data.

4-3 Design Loads

The size of the equipment should be based on the **Manual J** cooling loads for the entire house. Both the sensible and the latent loads are required. The **Manual J** heating load also will be required when the balance-point diagram is generated.

4-4 Air-source Heat Pump Sizing Limits

When heating and cooling is required, the heat pump equipment should be sized so that the sensible cooling capacity is greater than the calculated sensible load and the latent cooling capacity is greater than the latent load. Ideally, the total cooling capacity should not exceed the total cooling load by more than 15 percent. However, in colder climates, the total cooling capacity may exceed the total cooling load by as much as 25 percent. (A larger package will produce a lower thermal balance point and this will translate into lower operating costs during the heating season.)

4-5 Balance Point

The heating season balance point is equal to the outdoor temperature at which the heating capacity of the heat pump is equal to the heating load of the structure. The value for an air-source heat pump balance point can range from less than 30 °F to more than 40 °F, depending on the climate, the thermal efficiency of building envelope, and the size of the heat pump package.

- Low-range balance points are associated with hot climates because the heating load is relatively small compared to the cooling load. (In some cases, the balance point can be lower than the winter design temperature.)

- High-range balance points are associated with cold climates because the heating load tends to be much larger than the cooling load.

Model 25		
1,200 CFM		
Outdoor Temperature °F	Heating BTUH	Input (Watts)
60	42,200	3,735
50	37,100	3,485
40	29,500	3,230
30	23,800	3,080
20	21,200	2,725
10	19,700	2,490
0	15,800	2,255
-10	11,900	2,025
1) BTUH @ 70 °F air into coil .... (+2% @ 60; - 2% @ 80). 2) Heating capacity derated for defrost cycle (70 % RH). 3) Input power for compressor, add 400 Watts for blower.		

Table 4-1

4-6 Balance-point Diagram Construction

A balance-point diagram can be created by plotting the manufacturer's heating performance data and the envelope heat loss data on a piece of graph paper. The balance point occurs at the intersection of these two lines.

Step 1 — Graph the Heating Performance Data

Figure 4-1 is a graph of the heating performance data that is summarized by Table 4-1. In this case, there is a jog in the capacity line. This jog, which is referred to as a "defrost knee," indicates that the data accounts for the inefficiencies associated with the defrost cycle. (Heating capacity lines that are perfectly straight do not account for the capacity reductions caused by the defrost cycle.)

• Some manufacturers publish the defrost-adjusted capacity data in graphical form. In this case, a copying machine can be used to produce the required worksheet.

• Most manufacturers publish the heating capacity data in tabular form. In this case, tabular data can be plotted on a piece of graph paper. This data may or may not include a defrost cycle adjustment.

• Some manufacturers publish only the heating capacities that correspond to the two ARI rating points (17 °F and 47 °F). In this case, a graph can be created by drawing a straight line through these points. However, this graph will not account for the capacity reduction that is associated with the defrost cycle. (Refer to Section 8 of this manual for a method that can be used to adjust this performance line for the effect of the defrost cycle.)

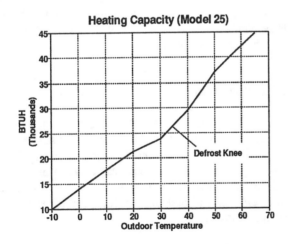

Figure 4-1

Step 2 — Graph the Performance of the Structure

Sketch the load line (heating load versus outdoor temperature) on the equipment performance graph. As indicated by Figure

4-2, the balance point occurs at the intersection of this line and the heating capacity line.

• Locate the point that corresponds to the **Manual J** winter design temperature and the **Manual J** heating load.

• Locate the point that corresponds to 65 °F and no heating load.

• Draw a straight line between these two points.

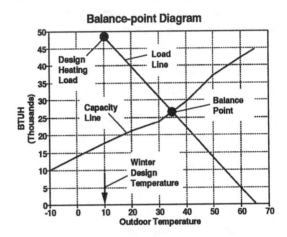

Figure 4-2

4-7 Optimum Balance Point

In cold climates, the electrical utility load measured in Kilowatts (Kw) and the heating season operating costs could be minimized if the heat pump package is selected to produce a low-range balance point. (A low balance point value means that more heat will be provided by the compression-cycle machinery and less heat will be provided by the resistance heater.) However, the heat pump package cannot be arbitrarily oversized; if the equipment is too large, the cooling performance will be degraded, particularly in regard to its ability to control the indoor humidity and the temperature gradients within the conditioned space. (Poor humidity control is not an important factor if the home is located in an arid climate.) Therefore, as far as the size of the heat pump equipment is concerned, the lowest legitimate balance point will be established by the 25 percent limit on excess cooling capacity.

If the lowest legitimate balance point is undesirably high, it can be lowered by increasing the **heating season** efficiency of the building envelope. For example, insulation can be added to reduce the heat losses that are associated with below-grade walls or ground slabs. (Select insulation and sealing options that have a major effect on the heating load and a marginal effect on the cooling load.)

Note that a lower balance point does not always translate into a significant reduction in operating costs. For example, Table 4-2 shows that a half-ton of extra cooling capacity reduces the balance point by only a few degrees. This figure also indicates that this shift translates into a $40 to $50 dollar annual savings in a cold climate city (like Akron, OH or Minneapolis, MN), and that the cost of operation could either increase or be marginally effected if the city is located in a warm climate or a moderate climate (Atlanta, GA or Seattle WA, for this example).

Also note that the cash flow benefit that is associated with a balance point reduction is closely related to the cost of the electrical power. For example, the annual savings associated with the Minneapolis home will be doubled if the utility rate is raised from 6 to 12 cents per kilowatt hour (KWH).

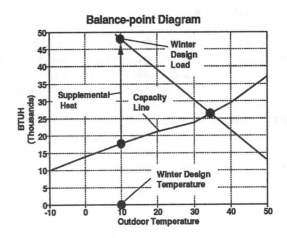

Figure 4-3

Akron, OH ($/KWH = 0.055)					
Capacity	% TC	Bal. Pt.	$ Heat	$ Cool	$ Total
2.5 Tons	1.15	34	1,026	166	1,192
3.0 Tons	1.40	31	978	170	1,148
Atlanta, GA ($/KWH = 0.055)					
Capacity	% TC	Bal. Pt.	$ Heat	$ Cool	$ Total
3.0 Tons	1.18	33	364	422	786
3.5 Tons	1.43	31	366	433	799
Minneapolis, MN ($/KWH = 0.060)					
Capacity	% TC	Bal. Pt.	$ Heat	$ Cool	$ Total
2.5 Tons	1.18	36	1624	192	1,816
3.0 Tons	1.40	33	1574	196	1,770
Seattle, WA ($/KWH = 0.060)					
Capacity	% TC	Bal. Pt.	$ Heat	$ Cool	$ Total
2.5 Tons	1.27	40	795	91	886
3.0 Tons	1.66	37	792	94	886

Table 4-2

4-9 Emergency Heat

Emergency heat is the total amount of resistance-coil heat that can be activated if the compressor fails. This heat can be provided by the supplemental heating coils, plus a reserve bank of heating coils that make up for the loss of the refrigeration cycle heating capacity. Figure 4-4 shows the relationship between emergency heat and supplemental heat.

Refer to local codes and utility regulations for information on the amount of heat that will be required if the compressor fails. If no codes or regulations apply, the total amount of failure-mode heating capacity will depend on the contractor's judgment and the owner's preference.

4-8 Supplemental Heat

Supplemental heat must be provided by electrical resistance coils when the outdoor temperature is below the balance point. As indicated by Figure 4-3, the maximum amount of supplemental heat that is required is equal to the difference between the winter design heating load and the capacity heat pump will have when it operates at the winter design temperature. This heat also is known as "second-stage heat" because it is activated by the second stage of the indoor thermostat. (An outdoor thermostat or an "intelligent thermostat" should be used to lock out the supplemental heat when the outdoor temperature is above the balance point.)

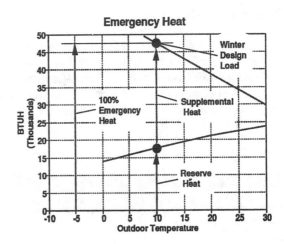

Figure 4-4

4-10 Reserve Heat

As indicated by Figure 4-5, the reserve heat is equal to the difference between the emergency heating requirement and the supplemental heating requirement. The purpose of the reserve heat is to compensate for the loss of refrigeration-cycle heating capacity. If possible, this heat should be provided by a bank of resistance coils that are locked out during normal operation. (In other words, the reserve coils are not activated by the second stage of the thermostat when the refrigeration machinery is operational.) Table 4-3 shows the desired control wiring.

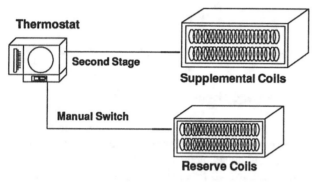

Figure 4-5

4-11 Auxiliary Heat

Auxiliary heat is the total amount of resistance-coil heat that is installed in the heat pump package. As indicated above, some of this heat is used to supplement the refrigeration cycle machinery during cold weather; and when the occasion arises, the remainder is used to compensate for malfunctioning machinery.

4-12 Mechanical Ventilation

If mechanical ventilation is required, the temperature of the air that enters the indoor coil will be cooler than the temperature of the return air (normally 70 °F). This effect is summarized by Table 1-3, which is repeated below. Note that these entering temperatures apply when the outdoor air CFM is equal to 10 percent of the blower CFM.

Approximate Entering Conditions — Indoor Coil 10 Percent Outdoor Air, 70 °F Return Air							
Outdoor db	-10	0	10	20	30	40	50
Entering db	62	63	64	65	66	67	68

Table 1-3 (Repeated)

4-13 Return-side Duct Loss

The temperature of the air entering the indoor coil is reduced by return-side conduction losses and leakage losses. This effect is summarized by Table 4-3. The top part of this table shows that the decrease in the temperature of the return air depends on the size of the conduction loss and the blower CFM. The bottom part of this table shows that, depending on the condition of the air that surrounds the return duct, a five to ten percent leakage rate will reduce the temperature of the return air by 1 °F to 8 °F. Also note that the temperature adjustments are additive when the conduction loss and the leakage loss occur simultaneously. (There are no losses associated with return ducts that are installed within the conditioned space.)

Return-side Duct Losses						
Conduction through Duct Wall						
Return Air Temperature Drop (°F) $$\frac{\text{Return-side Conduction Loss (BTUH)}}{1.1 \times \text{Blower CFM}}$$ See **Manual D**, Section 12-2 for information about conduction loss calculations.						
Return-side Leakage						
Percent of Blower CFM	**Return Air Temperature Drop (°F)**					
	Temperature of Air Surrounding Duct (°F)					
	-10	0	10	20	30	40	50
5 %	-4.0	-3.5	-3.0	-2.5	-2.0	-1.5	-1.0
10 %	-8.0	-7.0	-6.0	-5.0	-4.0	-3.0	-2.0
1) For leakage loss calculations, see **Manual D**, Section 12-3. 2) Refer to **Manual P**, for a procedure that can be used to determine the temperature of the return air for any amount of return-side leakage.						
Note						
When duct runs are installed in an unconditioned space, they should be sealed tightly and well insulated.						

Table 4-3

4-14 Air-to-Air Sizing Procedure

Initially, an air-to-air heat pump is sized to neutralize the cooling load. Then, after a product is selected, a balance point diagram is used to evaluate the heating performance. This diagram is also used to size the supplemental heating coils. The details of this procedure are summarized below.

Step 1 — Produce a Set of Design Parameters
Most of information that is required for sizing air-to-air heat pump equipment is generated by the **Manual J** procedure.

This basic collection of equipment sizing parameters includes the following items:

Cooling Season Data
- Outdoor dry-bulb temperature
- Indoor dry-bulb temperature
- Indoor wet-bulb temperature
- Sensible load from Procedure D
- Latent load from Procedure D

Heating Season Data
- Outdoor dry-bulb temperature
- Indoor dry-bulb temperature
- Sensible heating load
- Emergency heat requirement

If a ventilation load or a duct gain is associated with the return-side of the air distribution system, the condition of the air that enters the indoor coil will not be equal to the condition of the indoor air. When this is the case, Tables 1-2, 3-1 and 4-3 can be used to evaluate the condition of the air that approaches the indoor coil. This activity will produce three additional equipment sizing parameters:

- Entering dry-bulb temperature - cooling
- Entering wet-bulb temperature - cooling
- Entering dry-bulb temperature - heating

Step 2 — Size the Equipment for Cooling
Select a heat pump package that is compatible with the **Manual J** cooling requirements. Refer to Section 3 in this manual for a detailed discussion of this part of the equipment sizing procedure.

Step 3 — Evaluate the Heating Season Performance
If the balance point occurs at a relatively high temperature, a larger equipment package can be substituted for the package that is being evaluated. However, the total cooling capacity of the alternative package should not exceed the total load by more than 25 percent.

> As noted in Section 4-7, a larger heat pump may produce a lower balance point, but this extra capacity may not translate into a dramatic reduction in operating costs. In fact, the annual utility bill could increase if the climate is mild. And, even if the annual heating costs are reduced, the savings may not justify the cost of a larger unit. In other words, a comprehensive payback calculation should be used to validate the use of equipment that has excess cooling capacity.

Step 4 — Determine the Supplemental Heat Requirement
After the heat pump package is selected, use the balance-point diagram to determine how much supplemental heat will be required. The installed capacity should be equal to or slightly greater than the difference between the design heating load and the output of the heat pump when it operates at the winter design temperature.

Step 5 — Check the Emergency Heat Requirement
Compare the emergency heat requirement with the capacity of the supplemental heating coils. An additional bank of resistance heating coils will be needed if the emergency heat requirement exceeds the capacity of the supplemental coils.

Step 6 — Select a Resistance Heating Coil
Do not oversize the auxiliary heating coils (see Appendix A7). Refer to the manufacturer's performance tables and select a standard size coil that has a minimum amount of excess capacity. (Coils that can be activated in banks or stages offer more control options than single-stage coils.) Table 4-4 provides an example of resistance coil performance data.

Electric Heating Coil Performance				
Model	**BTUH**	**Kw**	**Stages**	**CFM***
HC-50	17,000	5.0	1	350
HC-75	25,600	7.5	1	525
HC-100-2	34,100	5.0 / 10.0	2	700
HC-125-1	42,700	12.5	1	900
HC-125-2	42,700	7.5 / 12.5	2	900
* Minimum CFM required to prevent overheating				

Table 4-4

4-15 Example — No Duct Loads, No Ventilation

This example shows how to use comprehensive performance data to size an air-to-air heat pump package. In this case the duct runs will be installed within the conditioned space and mechanical ventilation will not be required.

Step 1 — Design Data

Cooling
- Outdoor dry-bulb = 95 °F
- Indoor dry-bulb = 75 °F
- Indoor relative humidity = 50 percent
- Indoor wet-bulb = 63 °F
- Sensible load in conditioned space = 27,200 BTUH
- Latent load in conditioned space = 6,700 BTUH
- Total load = 33,900 BTUH

Heating
- Outdoor design temperature = 5 °F
- Indoor design temperature = 70 °F
- Heating load in conditioned space = 48,000 BTUH
- Emergency heat = **Manual J** heating load

Step 2 — Size the Equipment for Cooling

Initially, the equipment should be sized to meet the cooling requirements that are outlined in Step 1 above. Refer to Section 3-13 for the documentation that shows that the cooling loads can be satisfied by a Model 25 package (1,200 CFM, 4 percent excess capacity) or a Model 30 package (1,325 CFM, 20 percent excess capacity). Also refer to Tables 4-5 and 4-6 for the corresponding heating performance data.

Model 25		
1,200 CFM		
Outdoor Temperature °F	Heating BTUH	Input Watts
60	42,200	3,735
50	37,100	3,485
47	35,600	3,410
40	29,500	3,230
30	23,800	3,080
20	21,200	2,725
17	20,400	2,550
10	17,700	2,490
0	13,900	2,255
-10	10,000	1,725

1) BTUH @ 70 °F air into coil .... (+2% @ 60; - 2% @ 80).
2) Heating capacity derated for defrost cycle (70 % RH).
3) Input power for compressor, add 400 Watts for blower.

Table 4-5

Model 30		
1,325 CFM		
Outdoor Temperature °F	Heating BTUH	Input Watts
60	49,600	4435
50	43,500	4,120
47	41,600	4,025
40	34,400	3,795
30	27,500	3,480
20	24,100	3,170
17	23,100	3,075
10	20,000	2,890
0	15,700	2,620
-10	11,300	2,355

1) BTUH @ 70 °F air into coil .... (+2% @ 60; - 2% @ 80).
2) Heating capacity derated for defrost cycle (70 % RH).
3) Input power for compressor, add 500 Watts for blower.

Table 4-6

Step 3 — Evaluate the Heating Season Performance

Figure 4-6 is the balance-point diagram for this example. In this case, a 33 °F balance point is associated with the Model 25 package and a 30 °F balance point is associated with the Model 30 package.

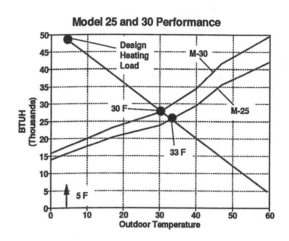

Figure 4-6

Step 4 — Determine the Supplemental Heat Requirement

Figure 4-7 shows that 10 Kw of supplemental heat will be required for either equipment package.(To convert supplemental BTUH values into Kw values, divide the BTUH values by 3,413.)

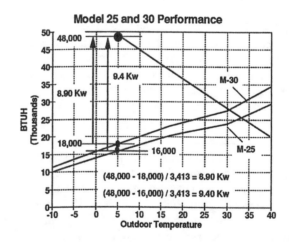

Figure 4-7

Step 5 — Check the Emergency Heat Requirement

A local regulation requires a 100 percent backup capability in the event of a compressor failure. This means that 14 Kw (48,000 BTUH) of resistance coil heat will be required.

Step 6 — Select a Resistance Heating Coil
Table 4-7 shows that the HC-150-3 coil can provide 15 Kw of coil capacity. Also note that capacity can be activated in three stages. (In this case the supplemental heat should be activated in two stages (5 Kw and 10 Kw), and the remaining capacity (5 Kw) can be held in reserve.

Electric Heating Coil Performance				
Model	BTUH	Kw	Stages	CFM*
HC-100-2	34,100	5.0 / 10.0	2	700
HC-125-1	42,700	12.5	1	900
HC-125-2	42,700	7.5 / 12.5	2	900
HC-150-1	51,200	15	1	1,100
HC-150-3	51,200	5/10/15	3	1,100
* Minimum CFM required to prevent excessive rise.				

Table 4-7

4-16 Example — Return-side Ventilation

This example is similar to the previous example, except it shows how to use a manufacturer's performance data to size air-to-air heat pump equipment when outdoor air is introduced through the return-side of the duct system. The values for the sensible and latent loads and the other sizing parameters that are associated with this example are summarized below. Also note that all the duct runs will be installed within the conditioned space. (The procedure in this example does not apply if the outdoor air CFM exceeds 10 percent of the blower CFM. Refer to **Manual P** for a procedure that can be used to evaluate the condition of the entering air if the outdoor air flow exceeds this 10 percent value.)

Step 1 — Design Parameters

Cooling
• Outdoor dry-bulb = 95 °F (humid climate)
• Indoor dry-bulb = 75 °F
• Indoor relative humidity = 50 percent
• Indoor wet-bulb = 63 °F
• Sensible load in conditioned space = 27,200 BTUH
• Latent load in conditioned space = 6,700 BTUH
• Ventilation rate = 100 CFM
• Sensible outdoor air load = 2,200 BTUH (Procedure D)
• Latent outdoor air load = 1,900 BTUH (Procedure D)
• Sensible design load = 29,400 BTUH (Procedure D)
• Latent design load = 8,600 BTUH (Procedure D)
• Total design load = 38,000 BTUH

Heating
• Outdoor design temperature = 5 °F
• Indoor design temperature = 70 °F
• Heating load in conditioned space= 48,000 BTUH

• Ventilation rate = 100 CFM
• Heating outdoor air load = 7,150 BTUH (J-Form, page 1)
• Heating design load = 55,150 BTUH (J-Form, page 1)
• Emergency heat = Heating design load

Step 2 — Size the Equipment for Cooling
Initially, the equipment should be sized to meet the cooling requirements that have been collected in Step 1 above. Refer to Section 3-15 for the documentation that shows that the cooling loads can be satisfied by a Model 30 package (1,325 CFM, 7 percent excess capacity). Also refer to Table 4-6, for the corresponding heating performance data.

Step 3 — Evaluate the Heating Season Performance
Since outdoor air is mixed with return air, the temperature of the air entering the indoor coil will be less than the temperature of the return air. This effect is illustrated by Table 1-3 (reproduced on page 4-4), which indicates that a 10 percent outdoor air fraction will translate into an entering air temperature that ranges from less than 62 °F to more than 68 °F. This in turn will produce an increase the output capacity of the heat pump, as indicated by the footnote that supplements the tabulated performance data (see Table 4-6). But, since the capacity increase is only a few percent, this effect can be ignored when the balance-point diagram is produced.

Figure 4-8 is the balance-point diagram for this example. To illustrate the comment made in the previous paragraph, this diagram shows that, for practical purposes, the balance point that is associated with the uncorrected capacity data and the balance point that is compatible with a 2 percent increase in the heating capacity are coincidental.

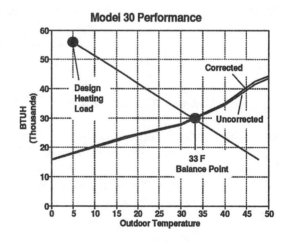

Figure 4-8

Step 4 — Determine the Supplemental Heat Requirement
On the next page, Figure 4-9 shows that 11 Kw of supplemental heat will be required for the Model 30 equipment package.

But since an 11 Kw heater is not standard equipment, the installed capacity of the supplemental heating coil may exceed the 11 Kw value.

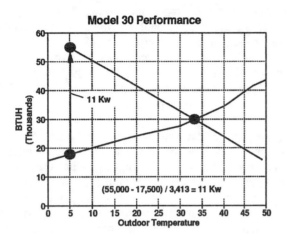

Figure 4-9

Step 5 — Check the Emergency Heat Requirement
A local regulation requires a 100 percent backup capability in the event of a compressor failure. This means that 16 Kw (55,000 BTUH) of resistance coil heat will be required.

Step 6 — Select a Resistance Heating Coil
Table 4-7 shows that the HC-150-3 coil can provide 15 Kw of heating capacity, which is 1 Kw short as far as the emergency heat requirement is concerned. However, this deficiency is not significant because it represents only a 2 percent shortage in emergency heating capacity. So this coil is acceptable on a technical basis, but a larger coil will be required to comply fully with the local regulation.

4-17 Example — Air-to-Air, Duct Loads

As noted in Section 3-16, duct runs that are installed in an unconditioned space must be sealed and insulated. This example shows how return-side duct losses affect the heating load, the performance of the heat pump package, the supplemental heating requirement and the emergency heating requirement.

Except for the location of the duct system (attic), this example is the same as the first example (Section 4-15, ducts in conditioned space). The values for the duct loads (cooling and heating) and the other sizing parameters associated with this example are summarized in step 1. Note that outdoor air is not used for mechanical ventilation, but there is an outdoor air load created by the return-side duct leakage. (See **Manual D** for information about duct gain calculations.)

Step 1 — Design Data

Cooling
- Outdoor dry-bulb = 95 °F (humid climate)
- Grains difference = 34 (**Manual J**, Table 1)
- Indoor dry-bulb = 75 °F
- Indoor relative humidity = 50 percent
- Indoor wet-bulb = 62-1/2 °F
- Sensible load in conditioned space = 27,200 BTUH
- Latent load in conditioned space = 6,700 BTUH
- Ventilation rate = no provision (0 CFM)
- Estimated conduction gains = 2,200 BTUH
- Estimated leakage = 80 CFM (supply), 100 CFM (return)

Heating
- Outdoor design temperature = 5 °F
- Indoor design temperature = 70 °F
- Heating load in conditioned space = 48,000 BTUH
- Ventilation rate = no provision (0 CFM)
- Duct installed in attic, 2-inch blanket
- Air in attic = 5 °F db
- Air off coil = 110 °F db (estimate)
- Estimated supply-side conduction loss = 1,560 BTUH
- Estimated supply-side leakage = 80 CFM
- Estimated return-side conduction loss = 2,420 BTUH
- Estimated return-side leakage = 100 CFM
- Emergency heat = **Manual J** heating load

Step 1A — Duct Leakage Load (Rounded)

Heating load = 1.1 x 80 x (110 - 5) = 9240 BTUH

Step 1B — Return-side Ventilation Effect

Heating load = 1.1 x (100 - 80) x (70 - 5) = 1430 BTUH

Step 1C — Equipment Sizing Load

Heat = 48000 + 1560 + 2420 + 9240 + 1430 = 62650 BTUH

Step 2 — Size the Equipment for Cooling
Initially, the heat pump package should be sized to neutralize the cooling loads. Refer to Section 3-16 for the documentation that shows these requirements can be satisfied by the Model 50 package. (Table 3-11, shows that when this unit operates at 1,600 CFM, it will have four percent excess capacity.) Also refer to Table 4-8, which can be found on the next page, for the heating-cycle performance table.

Step 3 — Evaluate the Heating Season Performance
On the next page, Figure 4-10 shows the balance-point diagram for the Model 50 package. In this case, the heating capacity line has not been adjusted for the effect of the return-side duct leakage. (The duct leakage causes the entering air temperature to be lower than 70 °F, but a 2 percent increase in the heating capacity will not have a noticeable effect on the balance point.)

Model 50		
1,600 CFM		
Outdoor Temperature °F	Heating BTUH	Input Watts
60	58,100	4,195
50	50,500	3,820
47	48,200	3,705
40	39,500	3,445
30	31,100	3,065
20	26,700	2,685
17	25,400	2,570
10	22,100	2,375
0	17,200	2,095
-10	12,400	1,820

1) Indoor temperature = 70 °F.... (+2% @ 65; - 2% @ 75).
2) Heating capacity derated for defrost cycle (70 % RH).
3) Input power for compressor, add 600 Watts for blower

Table 4-8

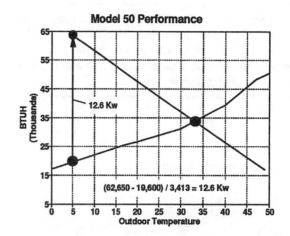

Figure 4-11

Step 6 — Select a Resistance Heating Coil
Table 4-9 shows that the HC-200-3 coil can provide 20 Kw of heating capacity. Also note that this coil can be activated in three stages. (In this case 15 Kw of supplemental heat can be activated by the two stages (7.5 and 15 Kw) and the remaining capacity (5 Kw) can be held in reserve.)

Electric Heating Coil Performance				
Model	BTUH	Kw	Stages	CFM*
HC-150-1	51,200	15	1	1,100
HC-150-3	51,200	5/10/15	3	1,100
HC-175-2	59,700	12.5/17.5	2	1,250
HC-175-3	59,700	5/12.5/17.5	3	1,250
HC-200-2	68,260	10/20	2	1,450
HC-200-3	68,260	7.5/15/20	3	1,450
* Minimum CFM required to prevent excessive rise.				

Table 4-9

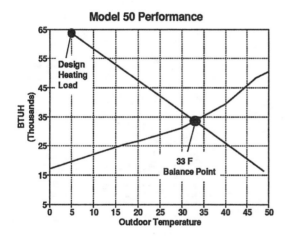

Figure 4-10

Step 4 — Determine the Supplemental Heat Requirement
Figure 4-11 shows that about 13 Kw of supplemental heat will be required. But since the heating coils are manufactured in standard sizes, the installed capacity of the resistance heating coils may exceed the 13 Kw value.

Step 5 — Check the Emergency Heat Requirement
The local regulation requires a 100 percent backup capability in the event of a compressor failure. Therefore, the total amount of emergency heating capacity is equal to 62,650 BTUH, which is equivalent to 18 Kw.

Table 4-10 compares this example with the example that featured duct runs within the conditioned space. This shows why energy codes try to discourage exposed duct runs.

	Exposed Duct	Not Exposed
Equipment	Model 50	Model 25
Total Cooling	48,200	35,400
CFM	1,600	1,200
Supplemental Heat	12.6 Kw	9.4 Kw
Emergency Heat	18 Kw	14 Kw

Table 4-10

It is possible to create an example problem that combines a ventilation load with a return-side duct leakage load. However, there is no difference between this problem and the previous example if the total flow of outdoor air is less than 10 percent of the blower CFM.

4-18 Heating-only Application

If cooling is not required, or if the summer climate is cool and dry, the size of the air-to-air heat pump equipment is not subject to the 25 percent oversizing limit. In this case, the equipment selection procedure can emphasize economic factors.

For example, if a 48,000 BTUH design load is associated with a 5 °F winter design temperature, a Model 25 or Model 30 package is suggested by the year-round design procedure (refer to Section 4-15). But if cooling performance is not important, a larger package could be installed and the heating costs will be correspondingly reduced. However, there are still constraints on the size of the equipment package. One of these constraints is technical and the other is economic.

- If compression-cycle heat is used to satisfy the entire design heating load (48,000 BTUH), an unacceptably large heat pump package will be required. For example, the four-ton, Model 50 package will deliver only about 20,000 BTUH when the outdoor temperature is equal to 5 °F (see Table 4-8). Note that this 20,000 BTUH heating capacity value is less than 50 percent of the design heating load.

- The costs of installing a larger package must be balanced against the reduction in operating costs. For example, if the energy costs associated with a five-ton package (2,250 CFM and 6 Kw of supplemental heat) are compared with the energy costs that are associated with the Model 25 package (1,200 CFM and 10 Kw of supplemental heat), the annual savings will be in the $100 to $200 range (in a cold climate at 6 cents/KWH). But since the differential in the installation cost will be between $1,500 and $2,000, it will take about 10 years to recover the investment in the larger equipment package.

In other words, when cooling is not a factor, the design can be based on an economic analysis, subject to "reasonable size" limitations. But this means that a comprehensive operation cost calculation must be used to justify equipment sizing decisions.

Section 5
Water-to-Air Heat Pump

5-1 Equipment Size Is Based on the Cooling Load

Water-to-air heat pumps normally provide year-round comfort, so the equipment selection procedure is identical to the procedure that would be used to select air-to-air equipment. This means that the size of the heat pump package must be compatible with the sensible and latent cooling loads and that a balance-point diagram will be required to evaluate the heating performance and the auxiliary heat requirement.

5-2 Manufacturer's Application Data

The equipment selection procedure requires comprehensive rating tables that specify the cooling performance associated with specific combinations of water-side equipment and indoor fan coil units. This performance data is published in a format that is similar to the example provided by Table 5-1.

W-36		Cooling Capacity Data			
		1,200 CFM			
EWT	GPM	Entering db / wb	Total Capacity	Sensible Capacity	Input Kw
50	4.5	75/63	38.7	27.6	2.67
		80/67	42.1	28.7	2.72
	7.0	75/63	39.9	28.1	2.57
		80/67	43.4	29.3	2.63

Table 5-1

W-36		Heating Capacity Data			
		1,200 CFM			
EWT	GPM	Entering db / wb	Heating Capacity	COP	Input Kw
50	7.0	60	37.9	3.75	2.96
		70	36.6	3.44	3.12
70	7.0	60	48.5	4.18	3.40
		70	47.0	3.85	3.58

Table 5-2

Tables that document the heating performance over a full range of entering water temperatures, water flow rates and entering air temperatures also are required. Table 5-2 provides

an example of this type of performance data. Notice that the heating capacity is not subject to a defrost penalty because water-to-air equipment does not have a coil that is exposed to the outdoor air.

Some manufacturers do not publish detailed equipment performance data, but it may be available upon request. If the appropriate capacity information is not available, the designer is advised to use a product supported by the necessary documentation. (Refer to Section 8 of this manual for an explanation of why the ARI rating data should not be used for equipment selection.)

5-3 Entering Water Temperature

As indicated by Tables 5-1 and 5-2, the cooling and heating capacity of a water-to-air heat pump depends on the temperature of the water that flows through the water-side of the equipment package. If this water is pumped from a well and then is discarded, the entering water temperature will be fairly constant throughout the year. But if the same water circulates through a closed piping-loop system that is buried in the ground, the temperature will change on a monthly basis, reaching a peak value (WTH) in the fall and a minimum value (WTL) in the spring.

The temperature excursions associated with a piping-loop system can be estimated by using the following equations, which require values for the hottest and coldest outdoor temperatures that are associated with the local climate. (Records that are maintained by local weather stations indicate that extreme temperatures seldom occur, but when they do they can be 10 °F to 20 °F hotter or colder than the **Manual J** design temperatures.)

WTH = Extreme summer temperature - 10 °F

WTL = Extreme winter temperature + 40 °F

Therefore, when a one-pass system is designed, a single water temperature can be used to evaluate the cooling performance and the heating performance. But if the water side of the system features a piping loop, the cooling performance must be evaluated at the highest water temperature value that could occur during the cooling season and the heating performance must be evaluated at the lowest water temperature value that could occur during the heating season.

Information about equipment performance is required for the work that is associated with designing a piping-loop. But since the WTH and WTL values are affected by the length of buried pipe and the circuit geometry, the piping-loop calculations must be based on the parameters that correspond to the estimated WTH and WTL values. Then, after the water loop is designed, the actual WTH and WTL values can be calculated. If these calculated values are significantly different than the estimated values, they can be used as the input for a second iteration of the piping-loop design procedure.

5-4 Design Loads

The size of the equipment should be based on the **Manual J** cooling loads for the entire house. Both the sensible and the latent loads are required. The **Manual J** heating load also will be required when the balance-point diagram is created.

5-5 Sizing Limitations

When heating and cooling is required, the heat pump equipment should be sized so that the sensible cooling capacity is greater than the calculated sensible load and the latent cooling capacity is greater than the latent load. Ideally, the total cooling capacity should not exceed the total cooling load by more than 15 percent. However, in colder climates, the total cooling capacity may exceed the total cooling load by as much as 25 percent (to obtain a lower thermal balance point and more heating efficiency).

5-6 Balance Point

The heating season balance point is equal to the outdoor temperature at which the heating capacity of the heat pump equals the heating load of the structure. The value for a water-source heat pump balance point can range from less than 20 °F to more than 30 °F, depending on the climate, the thermal efficiency of building envelope, the size of the heat pump package, and the design value for the entering water temperature.

• Low-range balance points are associated with hot climates because the heating load is relatively small compared to the cooling load.

• Low-range balance points are associated with hot climates because the entering water temperature (ground water or closed-loop) is relatively warm.

• Low-range balance points are associated with one-pass systems because the entering water temperature is always equal to the ground water temperature.

• High-range balance points are associated with cold climates because the heating load tends to be much larger than the cooling load.

• High-range balance points are associated with cold climates because the entering water temperature (ground water or closed-loop) is relatively cold.

• High-range balance points are associated with closed-loop systems because the entering water temperature can be much colder than the ground water temperature.

5-7 Balance-point Diagram Construction

A balance-point diagram can be created by plotting the manufacturer's heating performance data and the envelope heat loss data on a piece of graph paper. The balance point occurs at the intersection of these two lines.

Step 1 — Graph the Heating Performance Data
Figure 5-1 shows a graph of the heating performance data that is summarized by Table 5-2. In this case, the heating capacity is represented by a horizontal line because it is independent of the air temperature and because there is no defrost knee. This figure also shows how the heating capacity is affected by the entering water temperature.

• Some manufacturers publish the heating capacity data in 5 or 10 degree increments. Interpolation between the published rating points is recommended if the local water temperature is more than 2 degrees above or below the published value.

• Some manufacturers only publish the heating capacity for the 50 °F and 70 °F ARI rating points. In this case the heating performance can be estimated by interpolation between the two ARI rating points.

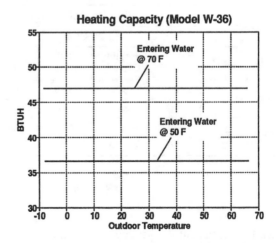

Figure 5-1

Step 2 — Graph the Performance of the Structure

Sketch the load line (heating load versus outdoor temperature) on the equipment performance graph. As indicated by Figure 5-2, the balance point occurs at the intersection of this line and the heating capacity line (70 °F water).

- Locate the point that corresponds to the **Manual J** winter design temperature and the **Manual J** heating load.

- Locate the point that corresponds to 65 °F and no heating load.

- Draw a straight line between these two points.

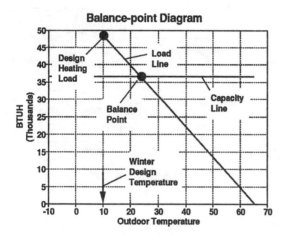

Figure 5-2

5-8 Optimum Balance Point

In cold climates, the electrical utility load (Kw) and the heating season operating cost could be minimized if the heat pump package is selected to produce a low-range balance point. (A low balance point value means that more heat will be provided by the compression-cycle machinery and less heat will be provided by the resistance heater.) However, the heat pump package cannot be oversized arbitrarily, if the equipment is too large, the cooling performance will be degraded, particularly in its ability to control the indoor humidity and the temperature gradients within the conditioned space. (Poor humidity control is not an important factor if the home is located in an arid climate.) Therefore, as far as the size of the heat pump equipment is concerned, the lowest legitimate balance point value will be established by the 25 percent limit on excess cooling capacity.

If the lowest legitimate balance point is undesirably high, it can be lowered by increasing the **heating season** efficiency of the building envelope. For example, insulation can be added to reduce the heat losses associated with below-grade walls or ground slabs. (Select insulation and sealing options

that have a major effect on the heating load and a marginal effect on the cooling load.)

Note that a lower balance point does not always translate into a significant reduction in operating costs. For example, Table 5-3 shows that a half-ton of extra cooling capacity reduces the balance point by only a few degrees. This figure also indicates that this shift translates into a $10 to $30 dollar annual savings in a cold climate city (Such as Akron, OH or Minneapolis, MN) and that the cost of operation could either increase or be marginally affected if the city is located in a warm climate or a moderate climate (Such as Atlanta, GA or Seattle, WA)

Also note that the cash flow benefit from a balance-point reduction is closely related to the cost of the electrical power. For example, the annual savings that are associated with the Minneapolis home will be doubled if the utility rate is raised from 6 to 12 cents per KWH.

Akron, OH ($/KWH = 0.055)					
Capacity	% TC	Bal. Pt.	$ Heat	$ Cool	$ Total
2.5 Tons	1.05	33	773	119	892
3.0 Tons	1.34	30	758	122	880
Atlanta, GA ($/KWH = 0.055)					
Capacity	% TC	Bal. Pt.	$ Heat	$ Cool	$ Total
2.5 Tons	1.23	32	311	326	637
3.0 Tons	1.35	28	320	331	651
Minneapolis, MN ($/KWH = 0.060)					
Capacity	% TC	Bal. Pt.	$ Heat	$ Cool	$ Total
2.5 Tons	1.03	35	1223	134	1,357
3.0 Tons	1.32	31	1195	138	1,333
Seattle, WA ($/KWH = 0.060)					
Capacity	% TC	Bal. Pt.	$ Heat	$ Cool	$ Total
2.5 Tons	1.14	33	623	75	698
3.0 Tons	1.59	30	639	78	717

Table 5-3

5-9 Supplemental Heat

Supplemental heat must be provided by resistance coils when the outdoor temperature is below the balance point. As indicated on the next page by Figure 5-4, the maximum amount of supplemental heat required is equal to the difference between the winter design heating load and the capacity that the heat pump will have when it operates at the winter design temperature. This heat also is known as "second-stage heat" because it is activated by the second stage of the indoor thermostat. (An outdoor thermostat or an "intelligent thermostat" should be used to lock out the supplemental heat when the outdoor temperature is above the balance point.)

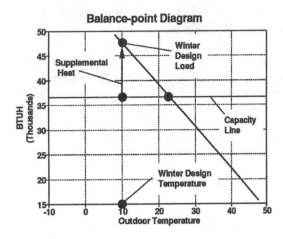

Figure 5-4

5-10 Emergency Heat

Emergency heat is the total amount of resistance-coil heat that can be activated if the compressor fails. This heat can be provided by the supplemental heating coils plus a reserve bank of heating coils that make up for the loss of the refrigeration cycle heating capacity. Figure 5-5 shows the relationship between emergency heat and supplemental heat.

Refer to local codes and utility regulations for information on the amount of heat that will be required if the compressor fails. If no codes or regulations apply, the total amount of failure-mode heating capacity will depend on the contractor's judgment and the owner's preference.

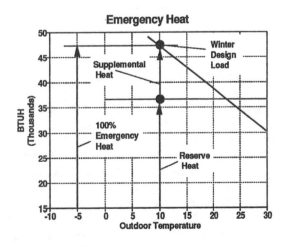

Figure 5-5

5-11 Reserve Heat

As indicated by Figure 5-5, the reserve heat equals the difference between the emergency heating requirement and the supplemental heating requirement. The purpose of the reserve heat is to compensate for the loss of refrigeration-cycle heating capacity. If possible, this heat should be provided by a bank of resistance coils that are locked out during normal operation. (In other words, the reserve coils are not activated by the second stage of the thermostat when the refrigeration machinery is operational.) Figure 5-6 shows the desired control wiring.

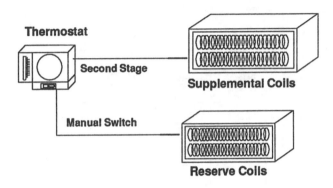

Figure 5-6

5-12 Auxiliary Heat

Auxiliary heat is the total amount of resistance-coil heat installed in the heat pump package. As indicated above, some of this heat is used to supplement the refrigeration-cycle machinery during cold weather; when the occasion arises, the remainder is used to compensate for malfunctioning machinery.

5-13 Mechanical Ventilation

If mechanical ventilation is required, the temperature of the air that enters the indoor coil will be cooler than the temperature of the return air (70 °F). This effect is summarized by Table 1-3, which is repeated below. Note that these entering temperatures apply when the outdoor air CFM is equal to 10 percent of the blower CFM.

Approximate Entering Conditions — Indoor Coil 10 Percent Outdoor Air, 70 °F Return Air							
Outdoor db	-10	0	10	20	30	40	50
Entering db	62	63	64	65	66	67	68

Table 1-3 (Repeated)

5-14 Return-side Duct Loss

The temperature of the air entering the indoor coil is reduced by return-side conduction losses and leakage losses. This effect is summarized by Table 4-3, which is repeated below. The top part of this table shows that the decrease in the temperature of the return air depends on the size of the conduction loss and the blower CFM. The bottom part of this table shows that, depending on the condition of the air that surrounds the return duct, a 5 to 10 percent leakage rate will reduce the temperature of the return air by 1 °F to 8 °F. Also note that the temperature adjustments are additive when the conduction loss and the leakage loss occur simultaneously. (There are no losses associated with return ducts that are installed within the conditioned space.)

Return-side Duct Losses							
Conduction through Duct Wall							
Return Air Temperature Drop (°F)							
$\dfrac{\text{Return-side Conduction Loss (BTUH)}}{1.1 \times \text{Blower CFM}}$							
See **Manual D**, Section 12-2 for information about conduction loss calculations							
Return-side Leakage							
Percent of Blower CFM	**Return Air Temperature Drop (°F)**						
	Temperature of Air Surrounding Duct (°F)						
	-10	0	10	20	30	40	50
5 %	-4.0	-3.5	-3.0	-2.5	-2.0	-1.5	-1.0
10 %	-8.0	-7.0	-6.0	-5.0	-4.0	-3.0	-2.0
1) For leakage loss calculations, see **Manual D**, Section 12-3. 2) Refer to **Manual P**, pages 23 and 24, for a procedure that can be used to determine the temperature of the return air for any amount of return-side leakage.							
Note							
When duct runs are installed in an unconditioned space, they should be sealed tightly and well insulated.							

Table 4-3 (Repeated)

5-15 Water-to-Air Sizing Procedure

Initially, water-to-air heat pumps are sized to neutralize the cooling load. Then, after a product is selected, a balance-point diagram can be used to evaluate the heating performance. This diagram is also used to size the supplemental heating coils. A step-by-step summary of this procedure is provided below.

Step 1 — Produce a Set of Design Parameters
Most of the information that is required for sizing water-to-air heat pump equipment is generated by the **Manual J** procedure. This basic collection of equipment sizing parameters includes the following items:

Cooling Season Data
- Outdoor dry-bulb temperature
- Indoor dry-bulb temperature
- Indoor wet-bulb temperature
- Sensible load from Procedure D
- Latent load from Procedure D

Heating Season Data
- Outdoor dry-bulb temperature
- Indoor dry-bulb temperature
- Sensible heating load
- Emergency heat requirement

If a ventilation load or a duct gain is associated with the return-side of the air distribution system, the condition of the air that enters the indoor coil will not be equal to the condition of the indoor air. When this is the case, Tables 1-2, 3-1 and 4-3 can be used to evaluate the condition of the air that approaches the cooling coil. This activity will produce three additional equipment sizing parameters:

- Entering dry-bulb temperature — cooling
- Entering wet-bulb temperature — cooling
- Entering dry-bulb temperature — heating

Since the capacity of water-to-air heat pump depends on the temperature of the water that flows through the equipment, an estimate of the entering water temperature will be required. Refer to Section 5-3 for more information about the following design parameters:

- Estimated entering water temperature — cooling
- Estimated entering water temperature — heating

The capacity of a water-to-air heat pump also depends on the water-side flow rate. If the design features a one-pass water system, the water flow rate is arbitrary, so the design can be based on the lowest flow rate listed in the performance data. (An inspection of the performance data shows that the capacity of the equipment is relatively insensitive to the water-side flow rate. So there is no reason to use any more water than is minimally required.) But if the design features a closed-loop piping system, a specific GPM value will be required to ensure turbulent flow in the piping circuit. Since this flow rate is usually robust, the equipment size should be based on one of the midrange flow rates that are listed in the performance data. (A maximum flow rate may be required when the water loop is actually designed, but this will not cause a problem because a larger flow rate value translates into a slight increase in equipment capacity.) Therefore, one of the following items must be included in the list of equipment sizing parameters:

- Design based on minimum flow rate (one-pass system)
- Design value for the water-side flow rate (loop system)

Step 2 — Size the Equipment for Cooling

Select a heat pump package that is compatible with the **Manual J** cooling requirements. Refer to Section 3 in this manual for a detailed discussion of this part of the equipment sizing procedure.

Step 3 — Evaluate the Heating Season Performance

If the balance point occurs at an undesirably high temperature, a larger equipment package can be substituted for the package that is being evaluated. However, the total cooling capacity of the alternative package should not exceed the total load by more than 25 percent.

Step 4 — Determine the Supplemental Heat Requirement

After the heat pump package is selected, use the balance-point diagram to determine how much supplemental heat will be required. The installed capacity should be equal to or slightly greater than the difference between the design heating load and the output of the heat pump when it operates at the winter design temperature.

Step 5 — Check the Emergency Heat Requirement

Compare the emergency heat requirement with the capacity of the supplemental heating coils. An additional bank of resistance heating coils will be needed if the emergency heat requirement exceeds the capacity of the supplemental coils.

Step 6 — Select a Resistance Heating Coil

Do not oversize the auxiliary heating coils. Refer to the manufacturer's performance tables and select a standard size coil that has a minimum amount of excess capacity. (Coils that can be activated in banks or stages offer more control options than single-stage coils.) Table 5-4 provides an example of resistance coil performance data.

Electric Heating Coil Performance				
Model	BTUH	Kw	Stages	CFM*
WHC-50	17,000	5.0	1	350
WHC-75	25,600	7.5	1	525
WHC-100-2	34,100	5.0 / 10.0	2	700
WHC-125-1	42,700	12.5	1	900
WHC-125-2	42,700	7.5 / 12.5	2	900
* Minimum CFM required to prevent over heating				

Table 5-4

5-16 Example — One-pass, No Return-side Loads

This example shows how to use comprehensive performance data to size a water-to-air heat pump package. In this case the duct runs will be installed within the conditioned space, mechanical ventilation will not be required and well water will make one pass through the water side of the system.

Step 1 — Design Data

Cooling
- Outdoor dry-bulb = 95 °F
- Indoor dry-bulb = 75 °F
- Indoor relative humidity = 50 percent
- Indoor wet-bulb = 63 °F
- Sensible load in conditioned space = 27,200 BTUH
- Latent load in conditioned space = 6,700 BTUH
- Total load = 33,900 BTUH

Heating
- Outdoor design temperature = 5 °F
- Indoor design temperature = 70 °F
- Heating load in conditioned space = 48,000 BTUH
- Emergency heat = **Manual J** heating load

Water-Side
- Temperature of ground water = 50 °F
- Entering water temperature = 50 °F
- Water-side flow rate = minimum

Step 2 — Size the Equipment for Cooling

Initially, the equipment should be sized to meet the cooling requirements outlined in Step 1 above. Refer to Section 3-17 for the documentation that shows the cooling loads can be satisfied by a W-36 package (1,200 CFM, 14 percent excess capacity). Also refer to Table 5-5, for the corresponding heating performance data.

W-36		Heating Capacity Data			
		1,200 CFM			
EWT	GPM	Entering db / wb	Heating Capacity	COP	Input Kw
50	4.5	60	36.2	3.69	2.87
		70	35.0	3.39	3.02
	7.0	60	37.9	3.75	2.96
		70	36.6	3.44	3.12
	9.0	60	39.5	3.83	3.02
		70	38.2	3.52	3.18

Table 5-5

Step 3 — Evaluate the Heating Season Performance

On the next page, Figure 5-6 shows the balance-point diagram for this example. In this case, a 22 °F balance point is associated with the W-36 package.

Step 4 — Determine the Supplemental Heat Requirement

On the next page, Figure 5-7 shows that 3.8 Kw of supplemental heat will be required. (To convert supplemental BTUH values into Kw values, divide the BTUH values by 3,413.)

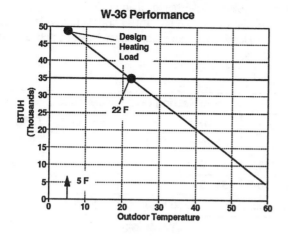

Figure 5-6

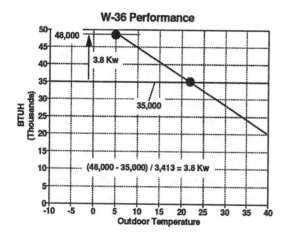

Figure 5-7

Step 5 — Check the Emergency Heat Requirement
A local regulation requires a 100 percent backup capability in the event of a compressor failure. This means that 14 Kw (48,000 BTUH) of resistance coil heat will be required.

Step 6 — Select a Resistance Heating Coil
Table 5-6 shows that the HC-150-3 coil can provide 15 Kw of coil capacity. Also note that capacity can be activated in three stages.

Electric Heating Coil Performance				
Model	BTUH	Kw	Stages	CFM*
HC-150-1	51,200	15	1	1,100
HC-150-3	51,200	5/10/15	3	1,100
* Minimum CFM required to prevent excessive rise				

Table 5-6

5-17 Example — Pipe Loop, No Return-side Loads

This example is identical to the previous example, except that the water-side of the system features a buried piping loop. This means that seasonal values for the entering water temperature will be required. This information is included in step 1, below (see the water-side data).

Step 1 — Design Data

Cooling
* Outdoor dry-bulb = 95 °F
* Indoor dry-bulb = 75 °F
* Indoor relative humidity = 50 percent
* Indoor wet-bulb = 63 °F
* Sensible load in conditioned space = 27,200 BTUH
* Latent load in conditioned space = 6,700 BTUH
* Total load = 33,900 BTUH

Heating
* Outdoor design temperature = 5 °F
* Indoor design temperature = 70 °F
* Heating load in conditioned space = 48,000 BTUH
* Emergency heat = **Manual J** heating load

Water Side
* Extreme temperature (summer) = 100 °F
* Extreme temperature (winter) = -10 °F
* WTH = 100 - 10 = 90 °F
* WTL = -10 + 40 = 30 °F
* Water-side flow rate determined by loop design

Step 2 — Size the Equipment for Cooling
Initially, the equipment should be sized to meet the cooling requirements that are outlined in Step 1. In this case, the cooling performance must be evaluated at a 90 °F entering water temperature value. Refer to Section 3-18 for the documentation that shows that the cooling loads can be satisfied by a W-41 package (1,375 CFM, 7 percent excess capacity). Also refer to Table 5-7, for the corresponding heating performance data (note the 30 °F water temperature).

W-41		Heating Capacity Data			
		1,375 CFM			
EWT	GPM	Entering db / wb	Heating Capacity	COP	Input Kw
30	5.0	60	32.7	3.27	2.93
		70	31.6	2.99	3.09
	8.0	60	34.5	3.34	3.03
		70	33.2	3.06	3.19
	11.0	60	35.7	3.39	3.09
		70	34.4	3.10	3.25

Table 5-7

Step 3 — Evaluate the Heating Season Performance
Figure 5-8 shows the balance-point diagram for this example. In this case, a 23 °F balance point is associated with the W-41 package when the entering water temperature (WTL) is equal to 30 °F.

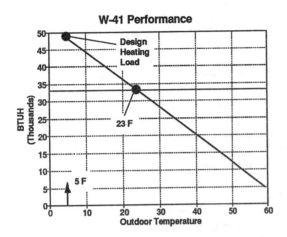

Figure 5-8

Step 4 — Determine the Supplemental Heat Requirement
Figure 5-9 shows that 5 Kw of supplemental heat will be required when the outdoor temperature is equal to the winter design temperature. (To convert supplemental BTUH values into Kw values, divide the BTUH values by 3,413.)

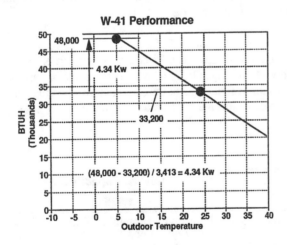

Figure 5-9

Step 5 — Check the Emergency Heat Requirement
A local regulation requires a 100 percent backup capability in the event of a compressor failure. This means that 14 Kw (48,000 BTUH) of resistance coil heat will be required.

Step 6 — Select a Resistance Heating Coil
Table 5-8 shows that the HC-150-3 coil can provide 15 Kw of coil capacity. Also note that capacity can be activated in three stages.

Electric Heating Coil Performance				
Model	**BTUH**	**Kw**	**Stages**	**CFM***
HC-125-2	42,700	7.5/12.5	2	900
HC-150-1	51,200	15	1	1,100
HC-150-3	51,200	5/10/15	3	1,100
* Minimum CFM required to prevent excessive rise				

Table 5-8

5-18 Example — One-pass with Ventilation

This example is similar to the Section 5-16 example (open water system), except it shows how to use a manufacturer's performance data to size water-to-air equipment when outdoor air is introduced through the return-side of the duct system. The values for the sensible and latent loads and the other sizing parameters that are associated with this example are summarized below. Also note that all the duct runs will be installed within the conditioned space. (The procedure presented by this example does not apply if the outdoor air CFM exceeds 10 percent of the blower CFM. Refer to **Manual P** for a procedure that can be used to evaluate the condition of the entering air if the outdoor air flow exceeds this 10 percent value.)

Step 1 — Design Parameters

Cooling
- Outdoor dry-bulb = 95 °F (humid climate)
- Indoor dry-bulb = 75 °F
- Indoor relative humidity = 50 percent
- Indoor wet-bulb = 63 °F
- Sensible load in conditioned space = 27,200 BTUH
- Latent load in conditioned space = 6,700 BTUH
- Ventilation rate = 100 CFM
- Sensible outdoor air load = 2,200 BTUH (Procedure D)
- Latent outdoor air load = 1,900 BTUH (Procedure D)
- Sensible design load = 29,400 BTUH (Procedure D)
- Latent design load = 8,600 BTUH (Procedure D)
- Total design load = 38,000 BTUH

Heating
- Outdoor design temperature = 5 °F
- Indoor design temperature = 70 °F
- Heating load in conditioned space = 48,000 BTUH
- Ventilation rate = 100 CFM
- Heating outdoor air load = 7,150 BTUH
- Heating design load = 55,150 BTUH (Procedure D)
- Emergency heat = heating design load

Step 2 — Size the Equipment for Cooling
Initially, the equipment should be sized to meet the cooling requirements that have been collected in Step 1 above. Refer to Section 3-20 for the documentation that shows that the cooling loads can be satisfied by a W-36 package (1,200 CFM, 6 percent excess capacity). Also refer to Table 5-9, for the corresponding heating performance data.

W-36		Heating Capacity Data			
		1,200 CFM			
EWT	GPM	Entering db / wb	Heating Capacity	COP	Input Kw
50	4.5	60	36.2	3.69	2.87
		70	35.0	3.39	3.02
	7.0	60	37.9	3.75	2.96
		70	36.6	3.44	3.12
	9.0	60	39.5	3.83	3.02
		70	38.2	3.52	3.18

Table 5-9

Step 3 — Evaluate the Heating Season Performance
Since outdoor air is mixed with return air, the temperature of the air entering the indoor coil will be less than the temperature of the return air. This is illustrated by Table 1-3 (reproduced on page 5-4), which indicates that a 10 percent outdoor air fraction will translate into an entering air temperature that ranges from less than 62 °F to more than 68 °F, depending the temperature of the outdoor air. This, in turn will produce an increase the output capacity of the heat pump, as indicated by Table 5-9. But since the capacity increase is only a few percent, this effect can be ignored when the balance-point diagram is produced. Figure 5-10 shows the balance-point diagram that is associated with a 70 °F entering air temperature.

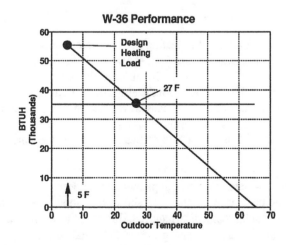

Figure 5-10

Step 4 — Determine the Supplemental Heat Requirement
Figure 5-11 shows that 5.9 Kw of supplemental heat will be required for the W-36 equipment package. In this case a 7-1/2 Kw heater would probably be the next largest standard size that can be found in manufacturers' catalogs.

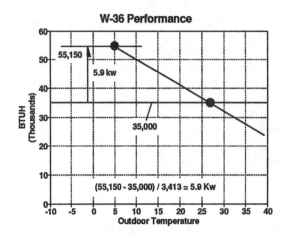

Figure 5-11

Step 5 — Check the Emergency Heat Requirement
A local regulation requires a 100 percent backup capability in the event of a compressor failure. This means that 16 Kw (55,000 BTUH) of electric resistance coil heat will be required.

Step 6 — Select a Resistance Heating Coil
Table 5-10 shows that the HC-150-3 coil can provide 15 Kw of heating capacity, which is 1 Kw short as far as the emergency heat requirement is concerned. However, this deficiency is not significant because it represents only a 2 percent shortage in emergency heating capacity. So this coil is acceptable on a technical basis, but a larger coil will be required to comply fully with the local regulation.

Electric Heating Coil Performance				
Model	BTUH	Kw	Stages	CFM*
HC-50	17,000	5.0	1	350
HC-75	25,600	7.5	1	525
HC-100-2	34,100	5.0 / 10.0	2	700
HC-125-1	42,700	12.5	1	900
HC-125-2	42,700	7.5 / 12.5	2	900
HC-150-1	51,200	15	1	1,100
HC-150-3	51,200	5/10/15	3	1,100
* Minimum CFM required to prevent over heating				

Table 5-10

If this example had featured a buried-loop piping system, the heating performance of a W-41 package would have been analyzed because the W-36 unit did not have enough cooling capacity (see the comment below Table 3-20 in Section 3-20). And the heating performance of the W-41 package would have been evaluated at an entering water temperature that is compatible with the local climate. This value could range from less than 30 °F in a cold climate to more than 60 °F in the warmest part of the country. Therefore, as far as heating performance is concerned, a buried-loop design could have less or more heating capacity than a similar one-pass design, depending on the difference between the ground water temperature and the WTL value that is associated with a buried piping loop. This heating capacity difference will translate into a lower or higher balance point and an increase or decrease in the supplemental heating requirement.

5-19 Water-to-Air with Duct Loads

Except for the location of the duct system (attic), this example is the same as the example in Section 5-16 (open water system, ducts in conditioned space). The values for the duct loads (heating and cooling) and the other sizing parameters that are associated with this example are summarized in step-1 below. Note that outdoor air is not used for mechanical ventilation, but there is an outdoor air load created by the return-side duct leakage. (See **Manual D** for information about duct gain calculations.)

Step 1 — Design Data

Cooling
- Outdoor dry-bulb = 95 °F (humid climate)
- Grains difference = 34 (**Manual J**, Table 1)
- Indoor dry-bulb = 75 °F
- Indoor relative humidity = 50 percent
- Indoor wet-bulb = 62-1/2 °F
- Sensible load in conditioned space = 27,200 BTUH
- Latent load in conditioned space = 6,700 BTUH
- Ventilation rate = no provision (0 CFM)
- Estimated conduction gains = 2,200 BTUH
- Estimated leakage = 80 CFM (supply), 100 CFM (return)

Heating
- Outdoor design temperature = 5 °F
- Indoor design temperature = 70 °F
- Heating load in conditioned space = 48,000 BTUH
- Ventilation rate = no provision (0 CFM)
- Duct installed in attic, 2-inch blanket
- Air in attic = 5 °F db
- Air off coil = 110 °F db (estimate)
- Estimated supply-side conduction loss = 1,560 BTUH
- Estimated supply-side leakage = 80 CFM
- Estimated return-side conduction loss = 2,420 BTUH

- Estimated return-side leakage = 100 CFM
- Emergency heat = **Manual J** heating load

Step 1A — Duct Leakage Load (Rounded)

Heating load = 1.1 x 80 x (110 - 5) = 9240 BTUH

Step 1B — Return-side Ventilation Effect

Heating load = 1.1 x (100 - 80) x (70 - 5) = 1430 BTUH

Step 1C — Equipment Sizing Load

48000 + 1560 + 2420 + 9240 + 1430 = 62650 BTUH

Step 2 — Size the Equipment for Cooling
Initially, the heat pump package should be sized to neutralize the cooling loads. Refer to Section 3-21 for the documentation that shows these requirements can be satisfied by the W-49 equipment package. (Table 3-21 shows that when this unit operates at 1,600 CFM, it will have 15 percent excess capacity.) Also refer to Table 5-11 for the heating-cycle performance table.

W-49		Heating Capacity Data			
		1,600 CFM			
EWT	GPM	Entering db / wb	Heating Capacity	COP	Input Kw
50	6.0	60	52.1	3.68	4.15
		70	50.4	3.38	4.37
	9.0	60	54.6	3.73	4.28
		70	52.7	3.43	4.51
	12.0	60	56.1	3.76	4.37
		70	54.2	3.45	4.60

Table 5-11

Step 3 — Evaluate the Heating Season Performance
On the next page, Figure 5-12 shows the balance-point diagram for the W-49 package. In this case, the heating capacity line has not been adjusted for the effect of the return-side duct leakage. (The duct leakage causes the entering air temperature to be lower than 70 °F, but a small increase in the heating capacity will not have a significant effect on the balance point.)

Step 4 — Determine the Supplemental Heat Requirement
On the next page, Figure 5-13 shows that about 4.2 Kw of supplemental heat will be required. But since the heating coils are manufactured in standard sizes, the installed capacity of the resistance heating coil will probably be equal to 5 Kw value.

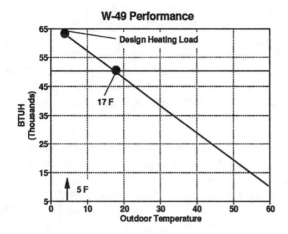

Figure 5-12

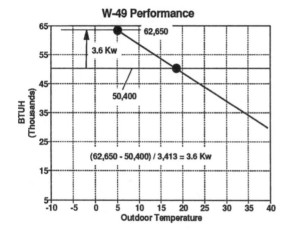

Figure 5-13

Electric Heating Coil Performance				
Model	**BTUH**	**Kw**	**Stages**	**CFM***
HC-175-2	59,700	12.5/17.5	2	1,250
HC-175-3	59,700	5/12.5/17.5	3	1,250
HC-200-2	68,260	10/20	2	1,450
HC-200-3	68,260	7.5/15/20	3	1,450
* Minimum CFM required to prevent excessive rise				

Table 5-12

Table 5-13 compares this example with the example that featured duct runs within the conditioned space. This shows why energy codes try to discourage exposed duct runs.

	Exposed Duct	**Not Exposed**
Equipment	W-49	W-36
Total Cooling	53,200	38,700
CFM	1,600	1,200
Supplemental Heat	3.6 Kw	5.1 Kw
Emergency Heat	18 Kw	14 Kw

Table 5-13

If this example had featured a buried-loop piping system, the heating performance of a W-49 package would have been analyzed at 12 GPM because the equipment was short of cooling capacity at 6 GPM (see the comment below Table 3-22 in Section 3-21). And the heating performance of the W-41 package would have been evaluated at an entering water temperature that is compatible with the local climate. This value could range from less than 30 °F in a cold climate to more than 60 °F in the warmest part of the country. Therefore, as far as heating performance is concerned, a buried-loop design could have less or more heating capacity than a similar one-pass design, depending on the difference between the ground water temperature and the WTH value that is associated with a buried piping loop. This heating capacity difference will translate into a lower or higher balance point and an increase or decrease in the supplemental heating requirement.

Step 5 — Check the Emergency Heat Requirement
The local regulation requires a 100 percent electric resistance backup capability in the event of a compressor failure. Therefore, the total amount of emergency heating capacity that will be required is equal to 62,650 BTUH, which is equivalent to 18 Kw.

Step 6 — Select a Resistance Heating Coil
Table 5-12 shows that the HC-200-3 coil can provide 20 Kw of heating capacity. Also note that this coil can be activated in three stages and that the first stage (5 Kw) has enough capacity to satisfy the supplemental heat requirement. (In this case the supplemental heat should be activated by the second stage of the thermostat and the remaining two stages should be off-line during normal operation. In the event that this extra capacity is needed, it can be energized by the emergency heating switch.)

5-20 Heating-only Heat Pump Sizing

Occasionally, cooling is not required or is provided by circulating the ground water directly through a water coil that is installed in a fan cabinet. If cooling is not required, or if the summer climate is cool and dry, the size of the water-to-air heat pump equipment is not subject to the 25 percent oversizing limit. In this case, the equipment selection procedure can emphasize economic factors.

For example, if a 48,000 BTUH design load is associated with a 5 °F winter design temperature, a W-36 Model (one-pass water system) or a W-41 model (buried-loop water system) is suggested by the year-round design procedure (refer to Sections 5-16 and 5-17). But if cooling performance is not important, a larger package could be installed and the heating season energy costs will be reduced. However, there are still technical and economic constraints on the size of the equipment package.

- If compression-cycle heat is used to satisfy the entire design heating load (48,000 BTUH), a large heat pump package may be required. In some cases the equipment may be unacceptably large. For example, if the design features a one-pass water system, the 5-ton W-49 Model will deliver about 50,000 BTUH when the entering water temperature is equal to 50 °F (see Table 5-11). However, this package will not satisfy the design heating load (48,000 BTUH) if a 30 °F entering water temperature is associated with a buried piping loop. (When the entering water temperature is equal to 30 °F, the heating capacity of the 5-ton model is about 39,000 BTUH.)

- The costs of installing a larger package must be balanced against the reduction in operating costs. For example, if the energy costs associated with a 5-ton package (1,600 CFM and no supplemental heat) are compared with the energy costs associated with the 3-1/2 ton package (1,200 CFM and 5 Kw of supplemental heat), the annual savings will be less than $15 per year (cold climate, 6 cents/KWH). Obviously, the cost associated with installing the larger package will not be justified by the reduction in the energy bill.

Section 6
Dual Fuel Systems

6-1 Overview

There are two ways to create a heating and cooling system that uses two fuels (usually natural gas and electricity). One method simply combines a split-system heat pump (air-to-air or water-to-air) with a conventional furnace. In this case, the furnace and the heat pump equipment can be provided by seperate manufactures. This system is similar to an electric-cooling-gas-heating system (see Figure 6-1).

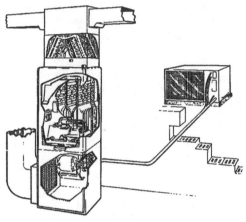

Figure 6-1

The other method consolidates the combustion equipment and the heat pump equipment into an integrated, split-system package. In this case, a single manufacturer is responsible for a design that features a burner located below the outdoor coil of an air-to-air heat pump (see Figure 6-2).

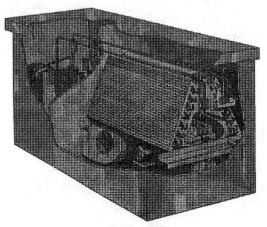

Figure 6-2

6-2 Concept

In some cases, depending on the cost of electricity and the price of the fossil fuel, it is less expensive to heat with compression-cycle equipment when the outdoor temperature is mild to moderately cold and more economical to heat with a fossil fuel when the outdoor temperature is very cold. When this is the case, the dual-fuel system will be less expensive to operate than a conventional heat pump system or a furnace-only system.

> Winter-peaking, electric utilities encourage home owners to install dual fuel systems because they translate into a reduced demand for power during periods of cold weather (heat provided by the furnace) and because they provide an opportunity to sell off-peak power during periods of moderately cold weather. This way, the electric utility optimizes its limited investment in generating capacity because severe weather conditions occur for relatively few hours per winter (compared to intermediate conditions). Understandably, the gas utilities are not enthusiastic about this concept.

6-3 Economic Balance Point

As noted in the previous paragraph, operational economics favor the heat pump equipment when the outdoor temperature is mild or moderately cold. But when the outdoor conditions are more severe, the furnace is the least expensive source of heat. So, it follows that there must be a single outdoor temperature associated with equal operating costs. This temperature is called the economic balance point.

6-4 Economic Balance-Point Diagram

On the next page, Figure 6-3 provides an example of an economic balance-point diagram. This diagram correlates the cost of delivering 1 million BTU of heat to the structure with the outdoor dry-bulb temperature. Notice that, as far as the furnace and the water source heat pump (WSHP) are concerned, the delivery cost holds fairly steady until the outdoor temperature rises above freezing, and then there is a only a slight increase in the delivery cost as the outdoor air gets warmer. (Cycling penalties start to have an effect when the outdoor temperature begins to moderate.) Also notice that an air-to-air heat pump is very sensitive to the temperature of the outdoor air, decreasing almost linearly with the increase in the outdoor temperature. (The shape and the relative positions of the performance curves vary with the cost of electricity and the cost of the fossil fuel.)

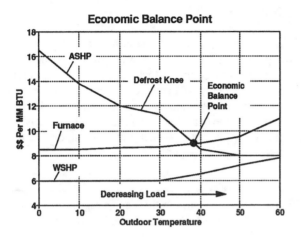

Figure 6-3

The economic balance point of a dual fuel system occurs at the intersection of the fossil fuel curve and the heat pump curve. In Figure 6-3, the economic balance point for the air-to-air heat pump is about 38 °F. Also notice that the water-source heat pump curve and the furnace curve do not intersect. This means simply that the delivery cost associated with the water-to-air equipment will always be less than the delivery cost associated with the furnace. (There is no economic balance point when one system is always the least expensive to operate.)

6-4 Break-even Coefficient of Performance

When operating in the heating mode, the fuel conversion efficiency of any type of heat pump equipment is defined by its coefficient of performance (COP). Since this parameter defines operational efficiency, it is related directly to the operating cost. This means that, given a set of fuel prices and a furnace efficiency value, there is a singular COP value associated with the economic balance point. In other words, there is a COP value that makes the cost of heating with the heat pump exactly equal to the cost of heating with the furnace. In this manual, this COP value will be called the "break-even coefficient of performance" (BECOP).

The BECOP value depends on the cost of electricity ($/KWH), the cost of the fossil fuel ($/1,000 Cu.Ft. for natural gas, $/Gallon for propane or oil) and the AFUE rating of the furnace. This value can be calculated by using one of the following formulas (constants evaluated in Section A1-9).

For Natural Gas:

$$BECOP = 293 \times AFUE \times \frac{\$/KWH}{\$/1000 \ CU.FT.}$$

For Oil:

$$BECOP = 41 \times AFUE \times \frac{\$/KWH}{\$/Gallon}$$

For Propane:

$$BECOP = 27 \times AFUE \times \frac{\$/KWH}{\$/Gallon}$$

Furnace AFUE values are published in the manufacturer's data and in the Gas Appliance Manufacturers Association (GAMA) directory. When published AFUE values are not available, the AFUE can be approximated by using the information that is provided by Table 6-1.

Approximate AFUE Values	
Gas Furnaces	**AFUE**
Atmospheric, draft hood, standing pilot	64.5
Atmospheric, direct draft, standing pilot, preheat	66.0
Atmospheric, draft hood, auto ignition	69.0
Forced, direct draft, auto ignition	75.0
Atmospheric, draft hood, auto ignition, vent damper	78.0
Induced draft, auto ignition, standard efficiency	78.0
Induced direct draft, auto ignition	78.0
Induced draft, auto ignition, high efficiency	81.5
Condensing	92.5
Note: 1) Draft hood — indoor combustion air with relief air 2) Direct draft — outdoor combustion air, no relief air 3) Induced draft — indoor combustion air, no relief air	
Oil Furnaces	**AFUE**
Standard design and efficiency	71.0
Standard design, improved efficiency	76.0
Standard design, improved efficiency, vent damper	83.0
Condensing	91.0

Table 6-1

6-5 Approximate Economic Balance Point

The break-even coefficient of performance value can be used to generate an approximate value for the economic balance point. This is accomplished by correlating the BECOP value with an operational COP value. However, the details of this calculation will vary, depending on the type of system.

Air-to-Air Heat Pump System

If air-to-air equipment is featured, there will be an economic balance point because the COP of the heat pump is very sensitive to changes in the outdoor air temperature. This relationship is defined by a COP versus outdoor air temperature graph, which will be similar (in appearance) to the example provided by Figure 6-4. This figure shows that if the BECOP value is equal to 2.23, the economic balance point will be equal to 31 °F.

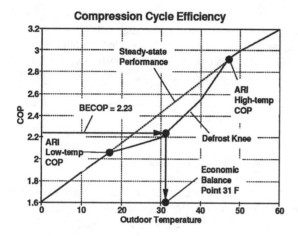

Figure 6-4

Unfortunately, manufacturers usually do not publish tables or graphs that show how the heating COP varies with outdoor temperature. However, they do publish data that correlates an output (BTUH) with a corresponding input (Kw or Watts) for a selected set of outdoor air temperatures. This information can be used to draw the COP versus OAT graph if one of the following equations is used to generate a set of COP values:

$$COP = \frac{Heating\ BTUH}{3.413 \times Watts}$$

$$COP = \frac{Heating\ BTUH}{3413 \times Kw}$$

For example, Figure 6-4 provides a graphical representation of the data and the calculations that are summarized by Table 6-2. Note that when the heating-mode output and input data is used to generate the COP versus outdoor air temperature data, the output values (BTUH) should include an allowance for the inefficiency associated with the defrost cycle. (Inspect Figure 6-4. Compare the steady-state performance line with the defrost knee line.) Also note that the input values (Kw) should include all of the electrical power that is consumed by the refrigeration-cycle equipment and the indoor blower.

Model 30 1,325 CFM				
Outdoor Temp.	Heating BTUH	Input [3] Watts	Blower Watts	COP Calc.
60	49,600	4,435	500	3.20
50	43,500	4,120	500	3.00
47	41,600	4,025	500	2.93
40	34,400	3,795	500	2.55
30	27,500	3,480	500	2.20
20	24,100	3,170	500	2.09
17	23,100	3,075	500	2.06
10	20,000	2,890	500	1.88
0	15,700	2,620	500	1.60

1) BTUH @ 70 °F air into coil .... (+2% @ 60; - 2% @ 80)
2) Heating capacity derated for defrost cycle (70 % RH)
3) Input power for compressor add 500 Watts for blower

Table 6-2

One-pass Water-to-Air System

If a water-to-air system features a one-pass piping circuit, the COP is essentially constant because the entering water temperature is not related to the outdoor air temperature. In this case, the BECOP value will either be larger or smaller than the operational COP value. If it is larger, the furnace will always be the most economical source of heat, so there is no justification for installing the heat pump. If it is smaller, the heat pump will always be the most economical source of heat, so the furnace would only be used as a supplemental heating device.

Water-loop Water-to-Air System

If a water-to-air system features a buried-loop piping circuit, the operational COP will change from month-to-month because the water-loop temperature cycles between a maximum value and a minimum value on a yearly basis. In this case, the economic balance point will be associated the entering water temperature value that produces an operational COP that is equal to the BECOP value. If this equivalency occurs, the heat pump will be more economical when the water temperature exceeds the economic balance point and the furnace will be more economical when the entering water temperature is below the economic balance point. However, it is quite possible to have a situation where the heat pump is always more efficient than the furnace. In this case, the furnace would only be used as a supplemental heating device.

6-6 Example — Economic Balance Point Calculation

When an air-to-air heat pump system is designed, the unit cost of electricity and natural gas are normally available. The furnace AFUE is usually known, or if not, it can be estimated by using the information that is provided by Table 6-1 on page 6-2.

For this example, assume that natural gas costs $7 per 1,000 Cu.Ft. and that electricity costs $0.065 per KWH and that the furnace AFUE is equal to 0.77. Then, determine the economic balance point if the performance of the heat pump is summarized by Table 6-3.

Air-to-Air Performance Data							
OAT	0	10	17	20	30	40	47
BTUH	13,900	17,700	20,400	22,200	24,800	29,500	35,600
Kw	2.65	2.89	2.95	3.12	3.42	3.63	3.81
Data Includes defrost adjustment and blower power							

Table 6-3

Step 1— Calculate the break-even COP

$$BECOP = 293 \times 0.77 \times \frac{0.065}{7.00} = 2.09$$

Step 2 — Use the manufacturer's performance data (Table 6-4) to calculate the COP values for a range of pertinent outdoor air temperatures.

COP Values							
OAT	0	10	17	20	30	40	47
COP	1.54	1.79	2.03	2.08	2.12	2.38	2.74
COP = BTUH / (3413 x Kw)							

Table 6-4

Step 3 — Compare the BECOP value (2.09) with the COP values (Table 6-4) to determine the outdoor air temperature that is associated with the economic balance point. In this case the economic balance point will occur at about 21 °F.

6-7 Investment Economics

Dual fuel systems are usually installed when the reduction in operating costs is large enough to justify the incremental increase in installation costs. In this regard, the operating and installation costs associated with a hybrid system should be compared with its most efficient competitors, which are the gas-heating-electric-cooling system and the conventional air-to-air or water-to-air heat pump system. (Comparisons with noncompetitive systems, such as an electric resistance heating system, make the return on investment appear deceptively attractive.)

Note that the economic justification for installing a dual fuel system is related to the economic balance point. In this regard, it is unlikely that an dual fuel system will be cost effective if the economic balance point exceeds 35 °F. However, no conclusions can be made about the potential return on investment if the economic balance point is less than 35 °F. In this case, a comprehensive energy use and operating cost calculation is required to demonstrate there is an economic advantage to installing a dual fuel system.

The economics of dual fuel systems can be significantly affected by utility incentive programs. For example, a payback calculation that indicates that the hybrid system is not cost effective when no incentives apply may be reversed if the utility offers a cash incentive or a "preferred system" rate schedule.

6-8 Furnace Sizing Procedure

The output capacity of the furnace must be equal to or greater than the **Manual J** design heating load, but preferably not more than 40 percent larger than the design heating load. In addition, the furnace must be equipped with a blower capable of delivering the air flow (CFM) required for the indoor refrigerant coil, and the blower must be able to develop enough pressure to overcome the resistance associated with the refrigerant coil and the remainder of the components in the air distribution system. In some cases, an oversized furnace may be required to obtain the desired blower performance. (Refer to Section 2 of this manual for a complete discussion of the furnace sizing guidelines.)

Quite often, a heat pump is added to an existing furnace. In this case, the air-side performance must be compatible with the heat pump equipment. This means that the existing blower must be capable of delivering the desired air flow when it operates against the resistance created by the refrigerant coil and the other components that are associated with the air distribution system. If the existing blower is not capable of producing the required air flow, a more powerful blower will have to be substituted for the existing blower or the air distribution system will have to be modified.

6-9 Heat Pump Sizing

The heat pump equipment should be sized so that the total cooling capacity is greater than the calculated load, but it should not exceed the total load by more than 25 percent. Also note that this sizing work should be based on the manufacturer's cooling performance data that corresponds to the **Manual J** summer design conditions. (Refer to Section 3 for a complete discussion of the sizing guidelines that pertain to cooling performance.)

After the heat pump equipment is sized to satisfy the sensible and latent cooling loads, it will be necessary check the heating performance. In this case the thermal balance point and the economic balance point must be evaluated. (The thermal balance point and the construction of the balance point diagram is discussed in Sections 4 and 5.)

6-10 Optimum Thermal Balance Point

The full economic benefit of an add-on heat pump will not be realized unless the thermal balance point is equal to or lower than the economic balance point. Therefore, the feasibility of using a larger heat pump package should be investigated when there is an adverse differential between the two balance points. However, the size of the heat pump package cannot be increased arbitrarily because severe oversizing may cause discomfort during the cooling season.

6-11 Retrofit Applications

If a new heat pump is added to an existing furnace, the blower will be required to produce an air flow that is compatible with the cooling loads and it will have to overcome the (considerable) air-side resistance created when the refrigerant coil is inserted into the air circulation path. There are two ways to check if the blower performance of an existing furnace will be acceptable after the refrigerant coil is installed in the supply duct.

Verification by Calculation
If the manufacturer's blower data is available, perform a **Manual J** load calculation and use this information and the blower data to generate a **Manual D** duct sizing calculation. Then check to see if the existing duct sizes are as large or larger than the sizes that are associated with the **Manual D** calculation.

Verification by Test
If the manufacturer's blower data is not available, use the field test method described below. Air balancing instruments are required for these tests. (A pitot tube and manometer will suffice.) It also will be necessary to bring along some sheets of cardboard. Proceed as follows:

• Lock out the furnace burner before beginning the test.

• Adjust the fan speed so that the blower is operating at its maximum speed.

• Simulate the pressure drop through a DX coil by blocking off a portion (about 50 percent) of the return filter with a piece of cardboard. Replace all covers and measure the pressure drop across the filter. Adjust the blockage until the filter section pressure drop is approximately equal to the pressure drop associated with a filter and a wet refrigerant coil. (The pressure drop across a standard flat filter or an electronic filter is normally less than 0.10 IWC and the pressure drop across a wet coil could range between 0.15 IWC and 0.30 IWC. The exact values can be found in the manufacturer's performance data.)

• Measure the supply CFM by making a pitot traverse of the main supply trunk. (If possible make this measurement about 6 feet downstream from the plenum).

Based on this test, the blower will be acceptable if it is able to deliver the desired air flow when the cardboard blockage (simulated coil) is in place. If the blower is unable to pass this test, the blower can be replaced by a more powerful blower, or some of the components in the duct system can be replaced with aerodynamically efficient components.

6-12 Example — Furnace and Air-to-Air Heat Pump

This example shows how to add an air-to-air heat pump to an existing gas furnace. In this case, the furnace has an input of 80,000 BTUH, an output 64,000 BTUH, an atmospheric burner, standing pilot, and no vent damper. No blower performance data is available. The home is in a locale where natural gas costs $ 7.50 per 1,000 cubic feet and electricity costs $0.075 per KWH.

Step 1 — Make a Load Calculation
A load calculation for the entire house will required to determine the equipment size. Room-by-room calculations also will be required because the performance of the existing duct system must be evaluated.

Step 2 — Summarize the Design Data

• Outdoor design db = 95 °F
• Room cooling dry-bulb = 75 °F
• Room cooling wet-bulb = 63 (75 °F, 50 percent RH)
• Sensible load = 27,200 BTUH
• Latent load = 6,700 BTUH
• Winter design temperature = 5 °F
• Heating load = 48,000 BTUH
• No outdoor air used for ventilation
• Duct runs in conditioned space

Step 3 — Size the Heat Pump for cooling
The heat pump equipment is sized to satisfy the cooling requirements outlined in Step 1 above. Refer to Section 3-13 for the documentation that shows that the cooling loads can be satisfied by a Model 25 package (1,200 CFM, 4 percent excess capacity) or a Model 30 package (1,325 CFM, 20 percent excess capacity). Also refer to Tables 4-5 and 4-6 for the corresponding heating performance data.

Step 4 — Determine the Thermal Balance Point
The manufacturer's integrated heating capacity data and the **Manual J** heating load information can be used to draw a thermal balance point diagram. In this case, Figure 6-5 shows that a 33 °F balance point is associated with the Model 25 package and a 30 °F balance point is associated with the Model 30 package.

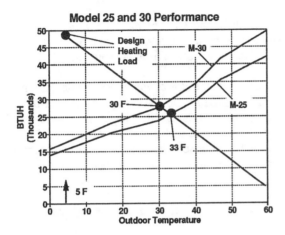

Figure 6-5

Step 5 — Estimate the Break-even COP
Since the furnace is older, it does not have an AFUE rating. However, Table 6-1 indicates that the AFUE is about equal to 64.5 percent, and, as indicated below, the BECOP value is equal to 1.89.

$$BECOP = 293 \times 0.645 \times \frac{0.075}{7.50} = 1.89$$

Step 6 — Estimate the Economic Balance Point
The output of the COP calculations are summarized by Table 6-5. Since the BECOP value is equal to 1.89, the economic balance point of the two candidate packages will be 17 °F or lower. (At this point, operation cost calculations would be required to select the package that offers the best return on investment. Also note that the potential for duct system problems increases with the size of the refrigerant coil CFM requirement.)

COP Calculations Model 25 and Model 30								
OAT °F	0	10	17	20	30	40	47	50
M-25	1.53	1.79	2.03	1.99	2.00	2.38	2.74	2.80
M-30	1.47	1.73	1.89	1.92	2.02	2.35	2.69	2.76

Table 6-5

Step 7 — Verify the Furnace Performance
The load calculation indicates that the furnace has considerably more heating capacity than is required, but adequate blower capacity is just as important. Since no blower data is available, a field test will be required to determine if the blower will be able to deliver the desired air flow rate after the DX cooling coil has been installed in the air path. (Refer to Section 6-11 for a summary of this test procedure.)

The blower will not be acceptable if, when tested at high speed, it is unable to deliver the required cooling CFM with the cardboard blockage (simulated coil) in place. If this is found to be the case, the blower or the duct system will have to be modified to obtain the desired air flow.

Step 8 — Verify Room Air Flow
The room loads and the **Manual D** duct sizing worksheets can be used to determine the air flow requirement for each room. If field tests indicate that a room is receiving an excessive or inadequate amount of air, the situation must be corrected. (If all of the duct runs are large enough, the system just needs to be balanced. If some duct runs produce too much air flow resistance, the sheet metal work will have to be modified.)

6-13 Example — Furnace and Water-to-Air Unit

This example shows how to add a water-to-air heat pump to an existing gas furnace. In this case, the furnace has an output capacity of 80,000 BTUH and a 0.815 AFUE rating. The home is in a city where gas costs $ 7.85 per 1,000 cubic feet and electricity costs $ 0.073 per KWH. Local ground water temperature is 50 °F. In this case the furnace blower performance data is available.

Step 1 — Make a Load Calculation
A load calculation for the entire house will required to determine the equipment size. Room-by-room calculations also will be required because the performance of the existing duct system must be evaluated.

Step 2 — Summarize the Design Data

- One-pass well water system
- Ground water temperature = 50 °F
- Room cooling dry-bulb = 75 °F
- Room cooling wet-bulb = 63 (75 °F, 50 percent RH)
- Sensible load = 27,200 BTUH
- Latent load = 6,700 BTUH
- Winter design temperature = 0 °F
- Heating load = 48,000 BTUH
- No outdoor air used for ventilation
- Duct runs in conditioned space

Step 3 — Size the Heat Pump for Cooling
The heat pump equipment is sized to satisfy the cooling requirements that are outlined in Step 1 above. Refer to

Section 3-17 for the documentation that shows that the cooling loads can be satisfied by a W-36 (1,200 CFM, 14 percent excess capacity). Also refer to Table 5-5, which indicates that the heat pump will deliver 35,000 BTUH and the compressor will draw 3.02 Kw when the entering water temperature is equal to 50 °F.

Step 4 — Estimate the Break-even COP
A value for the break-even coefficient of performance will be required to evaluate economic performance. As indicated below, the BECOP value is equal to 2.22.

$$BECOP = 293 \times 0.815 \times \frac{0.073}{7.85} = 2.22$$

Step 5 — Compare Operating Cost
There is no reason to install the water-to-air heat pump if the heat pump operating cost is expected to be more than the furnace operating cost. To make this determination, the 2.22 BECOP value can be compared to the coefficient of performance that the heat pump will have when the entering water temperature is 50 °F. The following calculations show that the water-to-air equipment will be the most economical source of heat because the operational COP exceeds the 2.22 benchmark.

- Heat output @ 50 °F = 35,000 BTUH
- Compressor power = 3.02 Kw
- Blower power = 400 Watts (from blower table, not shown)
- Pump power = 100 Watts (estimated)

$$\text{Operational COP} = \frac{35000}{3413 \times (3.02 + 0.4 + 0.1)} = 2.91$$

Step 6 — Verify the Furnace Performance
The load calculation indicates that the furnace has an excessive amount of heating capacity. And reference to the furnace

manufacturer's blower table indicates that, when operated at high speed, 1,200 CFM can be delivered against a resistance of 0.49 IWC. Therefore, the existing blower will be acceptable if the **Manual D** worksheets indicate that this is enough pressure to move 1,200 CFM through the refrigerant coil, filter, balancing damper, supply outlet, return grille, duct fittings, and the straight duct sections. (If the blower is inadequate, either the blower or the duct system will have to be modified to obtain the desired air flow.)

Step 7 — Verify Room Air Flow
The room loads and the **Manual D** duct sizing worksheets can be used to determine the air flow requirement for each room. If field tests indicate that a room is receiving an excessive or inadequate amount of air, the situation must be corrected. (If all of the duct runs are large enough, the system will just need to be balanced. If one or more duct of the runs produce too much air flow resistance, the sheet metal work will have to be modified.)

6-14 Integrated Dual Fuel Equipment

If an integrated dual fuel package is installed, the fossil fuel burner will be located below the outdoor refrigerant coil. In this configuration, compression-cycle heat will be adequate when the outdoor temperature is above the thermal balance point, and simultaneous operation of the refrigeration-cycle equipment and the fossil fuel burner will be required when the outdoor temperature is below the thermal balance point. This means that the equipment package must be selected so that the combined heat output (burner plus refrigerant cycle) is equal to or greater than the design heating load. However, this also means that, in some cases, the equipment could have an excessive amount of cooling capacity. When this situation occurs, smaller refrigeration-cycle equipment can be used, provided that the manufacturer offers the option of installing a small electric resistance heating coil in the indoor fan-coil cabinet.

Section 7
Variable- and Dual-speed Equipment

7-1 Overview

Some combinations of envelope designs and weather conditions produce a balance between the design cooling load and the design heating load. In these cases, a single-speed equipment package that is sized to neutralize the design-day cooling loads (sensible and latent) will also be compatible with the design heating load (and vice versa).

On the other hand, there are many homes that have a substantial mismatch between the design cooling load and the design heating load. In these cases, adjustable-speed equipment has an advantage over single-speed equipment because the cooling and heating capacities can be matched with the mode of operation.

The other advantages that are associated with variable-speed equipment pertain to comfort and efficiency. These performance characteristics are enhanced when the capacity can be modulated to match the load. Comfort also benefits from continuous air circulation. (These features are normally associated with relatively sophisticated commercial systems.)

The primary disadvantage of adjustable-speed equipment is that it is relatively expensive to install. Usually, the home owner will want to know if the marginal increase in the installation cost can be justified by the reduction in the operating cost. (A comparison should be made with a similar single-speed system.)

7-2 Cooling Performance

Figure 7-1 shows how the total capacity of adjustable-speed cooling equipment can shift between a minimum value and a maximum value (two-speed equipment) or modulate between a minimum value and a maximum value (variable-speed equipment). Note that if the design load is relatively large, the equipment will cycle or modulate between the minimum and maximum speed when the outdoor temperature is above 81 °F and it will cycle on and off when the outdoor temperature is below 81 °F. Also note that if the design load is relatively small, the equipment will cycle or modulate between the minimum and maximum speed when the outdoor temperature is above 89 °F and it will cycle on and off when the outdoor temperature is below 89 °F. (The Figure 7-1 balance point values are not generic. They must be evaluated on a case-by-case basis.)

As far as the comprehensive performance data is concerned, the manufacturer may publish only the information that pertains to maximum-speed operation. In this case, the data table

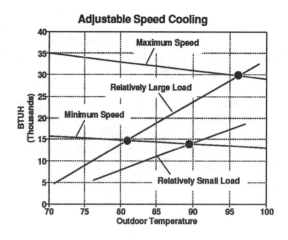

Figure 7-1

will be similar to the tables used to summarize the cooling-mode performance of single-speed equipment (see Tables 1-6 and 1-7).

Ideally, the manufacturer should provide two tables, one for maximum-speed operation and one for minimum-speed operation. In this case, both tables would be similar in appearance, but the capacity values and the power requirements would be noticeably different. (Information about the minimum-speed performance is useful for comparing the range of equipment capacity with the cooling load.)

7-3 Latent Capacity at Reduced Speed

Equipment that has an adequate amount of latent capacity when it operates at a high speed may not be able to control the indoor humidity when it operates at a reduced speed. (Evaporator temperatures depend on the mass flow of refrigerant, the surface area of the coil, and the air flow through the coil.) This behavior may not be noted in the manufacturer's performance data. (This is another reason for publishing the minimum-speed performance data.)

7-4 Sizing Adjustable-speed Cooling Equipment

With the exception of the limitation on excess cooling capacity, all of the guidelines, procedures and comments that are presented in Section 3 of this manual (single-speed cooling equipment) apply to adjustable-speed equipment. As far as the sizing limitation is concerned, the allowable margin of excess capacity will depend on the relative size of the design cooling load and the design heating load.

- If the cooling load is substantially larger than the heating load, the limit on excess cooling capacity is identical to the limit that applies to single-speed equipment (see Section 3-4).

- If the cooling load is substantially smaller than the heating load, an absolute limit on the amount of excess cooling capacity is not required because the equipment can be operated at a reduced capacity (speed reduction) during the cooling season. However, this does not mean that the amount of excess cooling capacity is irrelevant. In this regard, the designer must make sure that the system will provide adequate humidity control during any possible operating condition.

7-5 Heating Performance

Figure 7-2 shows how the heating capacity of a heat pump can shift between a minimum value and a maximum value (two-speed equipment) or modulate between a minimum value and a maximum value (variable-speed equipment). Note that if the design load is relatively large, the equipment will cycle on and off when the outdoor temperature is above 46 °F, it will modulate between the minimum and maximum speed when the outdoor temperature is between 46 °F and 35 °F, and it will operate continuously at a high speed when the outdoor temperature is below 35 °F. Also note that when the design load is relatively small, the equipment will cycle on and off when the outdoor temperature is above 37 °F, it will modulate between the minimum and maximum speed when the outdoor temperature is between 37 °F and 18 °F, and it will operate continuously at a high speed when the outdoor temperature is below 18 °F. (These balance-point values are not generic. They must be evaluated on a job-by-job basis.)

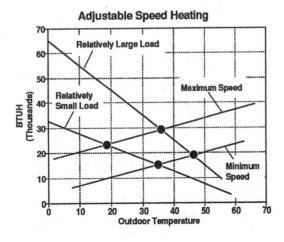

Adjustable Speed Heating

Figure 7-2

7-6 Balance-point Diagram

As far as the balance-point diagram is concerned, the procedures and comments presented in Section 4 of this manual (single-speed equipment) apply to adjustable-speed packages. However, as indicated in Section 7-4, the amount of excess cooling capacity (at high speed) may exceed the 25 percent limit on excess cooling capacity if acceptable humidity control can be provided during any load condition. Also note that if the manufacturer provides maximum and minimum speed performance data, the balance-point diagram will feature a maximum-speed capacity line, a minimum-speed capacity line and a heating-mode load line.

An example of this version of the balance-point diagram is provided by Figure 7-3. In this example, the heat pump will operate at low speed until the outdoor temperature drops below 47 °F. It will modulate between the minimum and maximum speed when the outdoor temperature is between 47 °F and 32 °F and it will operate continuously at a high speed when the outdoor temperature is below 32 °F.

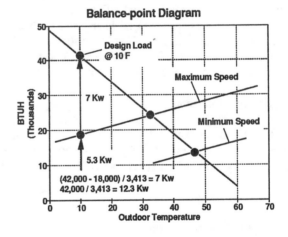

Balance-point Diagram

$(42,000 - 18,000) / 3,413 = 7 \text{ Kw}$
$42,000 / 3,413 = 12.3 \text{ Kw}$

Figure 7-3

7-7 Supplemental Heat

As with any type of heat pump, the capacity of the supplemental heating coils will be equal to the difference between the **Manual J** heating load and the high-speed output of the heat pump when the outdoor temperature is equal to the winter design temperature. (In the example illustrated by Figure 7-3, 7 Kw of supplemental heat will be required.)

7-8 Emergency Heat

Emergency heat is the total amount of resistance-coil heat that will be activated if the compressor fails. This heat can be provided by the supplemental heating coils plus a reserve bank of heating coils that make up for the loss of the refrigeration-cycle heating capacity. Figure 7-3 shows that 12.3 Kw (42,000 BTUH) of resistance heat will provide a 100 percent backup capability.

> Refer to local codes and utility regulations for information on the amount of heat that will be required if the compressor fails. If no codes or regulations apply, the total amount of failure-mode heating capacity will depend on the contractor's judgment and the owner's preference.

7-9 Reserve Heat

As indicated by Figure 7-3, the reserve heat is equal to the difference between the emergency heating requirement and the supplemental heating requirement. The purpose of the reserve heat is to compensate for the loss of refrigeration-cycle heating capacity. If possible, this heat should be provided by a bank of resistance coils that are locked out during normal operation. (In other words, the reserve coils are not activated by the second stage of the thermostat when the refrigeration machinery is operational.) Figure 7-3 shows that 5.3 Kw (18,000 BTUH) of resistance heat will compensate for the loss of refrigeration-cycle heating capacity.

7-10 Auxiliary Heat

Auxiliary heat is the total amount of resistance-coil heat installed in the heat pump package. As indicated above, some of this heat is used to supplement the refrigeration-cycle machinery during cold weather; and when the occasion arises,

the remainder is used to compensate for malfunctioning machinery.

7-11 Mechanical Ventilation

If mechanical ventilation is required, the condition of the air that enters the indoor coil will not be the same as the condition of the return air. This effect is summarized by Tables 1-2 and 1-3. Refer to Sections 3-15 and 4-16 for examples that show how to use this information.

7-12 Return-side Duct Loads

The condition of the air entering the indoor coil is affected by return-side conduction losses and leakage losses. These effects are summarized by Tables 3-1 and 4-3. Refer to Sections 3-16 and 4-17 for examples that show how to use this information.

7-13 Furnaces

Furnaces that have an adjustable-speed blower and a two-stage burner are desirable (and relatively affordable) when there is a significant difference between the size of the cooling load and the size of the heating load. If the cooling load is relatively large, a generous amount of blower capacity will be desirable during the cooling season and a reduction in blower capacity will be convenient during the heating season. (Reduced blower capacity will be compatible with the first stage of heating.) Or, if the heating load is relative large, the blower capacity can be reduced during the cooling season and modulated during the heating season (depending on which stage of burner capacity is activated).

Section 8
Limitations of the ARI Certification Data

8-1 ARI Certification Data

Not all manufacturers publish comprehensive application and sizing data, but virtually every manufacturer publishes ARI (Air-Conditioning and Refrigeration Institute) certification data. Unfortunately, this information is useful only for comparing the efficiency of similar pieces of refrigeration-cycle equipment that have been subjected to a specific set of test-stand conditions. The directory does not provide the type of information that is required for selecting and sizing refrigeration-cycle equipment.

8-2 Required Performance Parameters

Table 8-1 compares the information that is required for selecting and sizing refrigeration-cycle equipment with the information that is published in the ARI directory. Note that most of the data that is required for the design work that was discussed in the previous sections of this manual is not found in the directory.

Air-to-Air equipment				
Design Parameter	Required Data Points		ARI Rating Points	
	Cooling	Heating	Cooling	Heating
Indoor db °F	75	70	80	70
Indoor wb °F	60 to 66	NA	67	NA
Outdoor db °F	80 to 105	-20 to 50	95	47

Water-to-Air Equipment				
Design Parameter	Required Data Points		ARI Rating Points	
	Cooling	Heating	Cooling	Heating
Indoor db °F	75	70	80	70
Indoor wb °F	60 to 66	NA	67	NA
Water °F	30 to 100		50 and 70	

Table 8-2

Cooling Performance	
Required Information	ARI Directory
Nominal CFM	Not published
Total capacity (BTUH)	Total capacity (BTUH)
Sensible capacity (BTUH)	Not published
Latent capacity (BTUH)	Not published
Blower Table	Not published

Air-to-Air Heat Pump — Heating Performance	
Required Information	ARI Directory
Capacity versus OAT graph	Capacity at 47 °F
Defrost cycle adjustment	No adjustment

Table 8-1

In addition, the data that is published in the ARI directory is not based on **Manual J** indoor design conditions and it may not be compatible with the outdoor design temperature or the entering water temperature that is associated with a particular building site or a specific design. Table 8-2 summarizes the differences between the applied operating conditions and the ARI test conditions.

8-2 Range of Performance Data

A study has been made of the performance characteristics of a few hundred air-to-air equipment packages (cooling units and heat pumps) and about 30 water-to-air heat pumps. This information was found in the performance data that was published by four equipment manufacturers. The objective of the study was to:

- Find the range of CFM per Ton values that are used for the certification tests that were conducted in accordance with the ARI standards (210, 240, and 325).

- Find the range of "applied" CFM per Ton values that are authorized by the manufacturer's application data.

- Determine the approximate sensible-capacity-total-capacity ratio (S/T ratio) that is associated with various CFM per Ton air flow rates when the temperature of the air that is entering the indoor coil is equal to 75 °F db and 64 °F wb.

The results of this study are illustrated by three scatter plots that show the approximate range of these performance parameters. The first diagram demonstrates that the test-stand air flow rate is not necessarily equal to 400 CFM per Ton (see Figure 8-1 on the next page). This diagram also shows that the test-stand air flow values can range from 320 to 450 CFM per

Ton. (Note that the ARI standards do not specify that the tests must be conducted at a specific air flow rate; they only stipulate that the maximum flow rate be less than 450 CFM per Ton.)

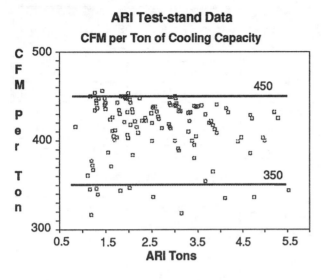

Figure 8-1

Figure 8-2 provides a diagram that shows the range of CFM/Ton values that are authorized by the manufacturer's application data. Notice that these values have a much wider range than the test-stand values (300 to 700 CFM per Ton, compared to 320 to 450 CFM per Ton). This means that a high rate of air flow can be used for applications that have a small latent load and a low rate of air flow can be used when the dehumidification load is relatively large.

Figure 8-3 presents the diagram that shows the range of S/T ratios that are associated with the CFM per Ton air flow rates that were authorized by the manufacturer's application data. (These ratios were calculated by dividing the sensible capacity that is associated with an entering air condition of 75 °F and 55 percent relative humidity by the total capacity value that is associated with the ARI rating.) For reference, a regression line that represents the best fit for the mean S/T ratio has been superimposed over the data points. Note that the S/T ratio that is associated with a particular equipment package can be considerably higher or lower than the mean value. This means that CFM per Ton data cannot be used to make an accurate estimate of the S/T ratio that is associated with a particular equipment package.

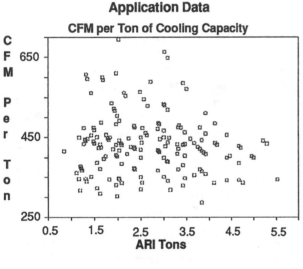

Figure 8-2

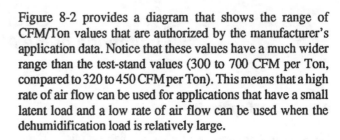
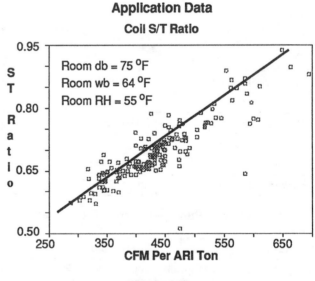

Figure 8-3

APPENDIX 1
Summary of Tables and Equations

This appendix collects all the primary tables and equations that were used in this manual. Refer to the original sections for instructions on how to use this information.

A1-1 Entering Conditions

Table 1-1 summarizes the entering conditions (cooling mode) when there is no mechanical ventilation or duct leakage. Equipment size should be based on performance tables that match the condition that is expected to occur at the site.

Entering Conditions — Cooling Coil No Mechanical Ventilation, No Return-side Duct Gains		
Relative Humidity	Entering db	Entering wb
55 Percent	75	64
50 Percent	75	62-1/2 (Use 63)
45 Percent	75	61-1/2 (Use 62)

Table 1-1

Table 1-2 summarizes the entering conditions (cooling mode) when there is mechanical ventilation (10 percent of the blower CFM), but no duct leakage. Equipment size should be based on performance tables that match the condition that is expected to occur at the site.

Approximate Entering Conditions — Cooling Coil 10 Percent Outdoor Air, No Return-side Duct Gains				
Relative Humidity	Humid Climate		Dry Climate	
	Entering db	Entering wb	Entering db	Entering wb
55 Percent	77	65-1/2	77	64
50 Percent	77	64	77	62-1/2
45 Percent	77	62-1/2	77	61
Refer to **Manual P**, for a procedure that can be used to determine precisely the condition of a mixture of return air and outdoor air.				

Table 1-2

Table 1-3 summarizes the entering conditions (heating mode) when there is mechanical ventilation (10 percent of the blower CFM), but no duct leakage. Equipment size should be based on performance tables that match the condition that is expected to occur at the site.

Approximate Entering Conditions — Indoor Coil 10 Percent Outdoor Air, 70 °F Return Air							
Outdoor db	-10	0	10	20	30	40	50
Entering db	62	63	64	65	66	67	68

Table 1-3

The following equation can be used to generate the entering dry-bulb (EDB) information that is presented by Table 1-3, and it also can be used to generate EDB values for any combination of outdoor dry-bulb temperature (ODB) and outdoor air percentage (OAP).

$$EDB = OAP \times ODB + 70 \times (1\text{-}OAP)$$

For example, a 62 °F entering dry-bulb temperature is associated with a 15 °F outdoor temperature and a 15 percent outdoor air fraction.

$$EDB = 0.15 \times 15 + 70 \times (1.00 - 0.15) = 61.75 = 62\ ^{\circ}F$$

A1-2 Sensible Heat Ratio Equation

Use the following equation to determine the sensible heat ratio that is associated with the **Manual J** loads. This ratio is useful because it can be used (see below) to establish a tentative value for the difference between the room temperature and the supply air temperature.

$$SHR = \frac{Manual\ J\ Sensible\ Load}{Manual\ J\ Total\ Load}$$

A1-3 Cooling TD

Use Table 1-4 to estimate the difference between the room temperature and the cooling supply air temperature.

Sensible Heat Ratio versus TD Value			
SHR	LAT	Room db	TD
Below 0.80	54	75	21
0.80 to 0.85	56	75	19
Above 0.85	58	75	17

Table 1-4

A1-4 Cooling CFM

Use the sensible heat equation (below) to calculate a preliminary value (candidate search value) for the cooling CFM.

$$CFM = \frac{Manual\ J\ Sensible\ Load}{1.1 \times TD}$$

A1-5 Summary of Equipment Selection Data

The following information is required for the equipment selection process. This list shows that sizing decisions should be based on performance tables that contain CFM, sensible capacity, latent capacity, dry-bulb and wet-bulb information.

Design Loads	Outdoor Conditions
Sensible	Summer dry-bulb
Latent	Summer wet-bulb
Heating	Winter dry-bulb
Room Conditions	**Air Entering Indoor Coil**
Dry-bulb — Cooling	Dry-bulb — cooling
Relative humidity	Wet-bulb — cooling
Dry-bulb — heating	Dry-bulb — heating
Air Flow Estimate	**Water Temperature**
TD (from Table 1-1)	Late summer
CFM (from equation)	Late winter

Table 1-5

A1-6 Limits on Furnace Capacity

Table 2-1 summarizes the sizing limits that apply to furnaces and boilers.

Limits on Excess Heating Capacity		
Criterion	Furnace	Boiler
Heating Comfort	40 %	40 %
Cooling Comfort	May exceed 40%	NA
Heating Efficiency	100 %	100 %

Table 2-1

A1-7 Temperature Rise

Use the sensible heat equation (top of next column) to estimate the furnace temperature rise.

$$Rise\ (^\circ F) = \frac{Output\ Capacity\ (BTUH)}{1.1 \times Heating\ CFM}$$

A1-8 Approximate AFUE Values

The "Approximate AFUE" table below was extracted from a similar table published in the 1988 ASHRAE Equipment Handbook. This table can be used to estimate the furnace AFUE rating when the published AFUE rating is not available. (Refer to the GAMA directory or the manufacturer's performance data for AFUE information.)

Approximate AFUE Values	
Gas Furnaces	**AFUE**
Atmospheric, draft hood, standing pilot	64.5
Atmospheric, direct draft, standing pilot, preheat	66.0
Atmospheric, draft hood, auto ignition	69.0
Forced, direct draft, auto ignition	75.0
Atmospheric, draft hood, auto ignition, vent damper	78.0
Induced draft, auto ignition, standard efficiency	78.0
Induced direct draft, auto ignition	78.0
Induced draft, auto ignition, high efficiency	81.5
Condensing	92.5
Note: 1) Draft hood — indoor combustion air with relief air 2) Direct draft — outdoor combustion air, no relief air 3) Induced draft — indoor combustion air, no relief air	
Oil Furnaces	**AFUE**
Standard design and efficiency	71.0
Standard design, improved efficiency	76.0
Standard design, improved efficiency, vent damper	83.0
Condensing	91.0

Table 6-1

A1-9 Heating Mode COP

Use the following equations to calculate the heating mode COP (heat pump equipment).

$$COP = \frac{Heating\ BTUH}{3.413 \times Watts}$$

$$COP = \frac{Heating\ BTUH}{3413 \times Kw}$$

A1-10 Break-even COP

Use the following equations to estimate the add-on heat pump break-even coefficient of performance (BECOP).

For Natural Gas:

$$BECOP = 293 \times AFUE \times \frac{\$/KWH}{\$/1000 \ Cu.Ft.}$$

$$293 = \frac{1,000,000 \ BTU \ per \ 1000 \ Cu.Ft.}{3,413 \ BTU \ per \ KWH}$$

For Oil:

$$BECOP = 41 \times AFUE \times \frac{\$/KWH}{\$/Gallon}$$

$$41 = \frac{140,000 \ BTU \ per \ gallon}{3,413 \ BTU \ per \ KWH}$$

For Propane:

$$BECOP = 27 \times AFUE \times \frac{\$/KWH}{\$/Gallon}$$

$$27 = \frac{92,000 \ BTU \ per \ gallon}{3,413 \ BTU \ per \ KWH}$$

A1-11 Return-side Duct Losses

Use Table 4-3 to estimate the effect that return-side duct losses have on the condition of the air entering the coil.

A1-12 Return-side Duct Gains

Use Table 3-1 to estimate the effect that return-side duct gains have on the condition of the air entering the coil.

Return-side Duct Losses

Conduction through Duct Wall

Return Air Temperature Drop (°F)

$$\frac{\text{Return-side Conduction Loss (BTUH)}}{1.1 \times \text{Blower CFM}}$$

See **Manual D**, Section 12-2 for information about conduction loss calculations

Return-side Leakage

Percent of Blower CFM	Return Air Temperature Drop (°F)						
	Temperature of Air Surrounding Duct (°F)						
	-10	0	10	20	30	40	50
5 %	-4.0	-3.5	-3.0	-2.5	-2.0	-1.5	-1.0
10 %	-8.0	-7.0	-6.0	-5.0	-4.0	-3.0	-2.0

1) For leakage loss calculations, see **Manual D**, section 12-3.
2) Refer to **Manual P** for a procedure that can be used to determine the temperature of the return air for any amount of return-side leakage.

Note

When duct runs are installed in an unconditioned space, they should be sealed tightly and well insulated.

Table 4-3

Return-side Duct Gains

Conduction through Duct Wall

Approximate Effect on the Condition of the Return Air	Sensible Gain (Percentage of the Sensible Load)			
	5%	10%	15%	20%
Return db	+1	+2	+3	+4
Return wb	+1/3	+2/3	+1	+1-1/3

1) For conduction loss calculations, see **Manual D**, Section 12-2.
2) Dry-bulb rise = (Return-side Duct Gain) / (1.1 x Blower CFM)
3) Wet-bulb rise can be read from the Psychrometric Chart.

Return-side Leakage (10 Percent)

Temperature of the Air Surrounding the Return Duct	Effect on Condition of Return Air			
	Humid Climate		Dry Climate	
	Entering db	Entering wb	Entering db	Entering wb
95 °F	+2	+1-1/2	+2	0
115 °F	+4	+2	+4	+1/2
135 °F	+6	+2-1/2	+6	+1

1) For leakage loss calculations, see **Manual D**, Section 12-3.
2) Refer to **Manual P** for a procedure that can be used to determine the condition of the return-air-ambient-air mixture for any amount of return-side leakage.

Note

When duct runs are installed in an unconditioned space, they should be sealed tightly and well insulated. This effort will yield two important benefits:

- The equipment size will be reduced because the sensible and latent loads associated with the duct runs will be minimized.

- It will be easier to size the cooling equipment because the dry-bulb and wet-bulb temperatures of the air that enters the cooling coil will be within 1 °F of the temperatures associated with the air returned from the conditioned space.

Table 3-1

APPENDIX 2
Manufacturer's Application Data

A2-1 Presentation of Data

Each manufacturer that publishes comprehensive equipment performance data does so in a slightly different format. The following pages in this appendix provide some examples of manufacturer's application data.

A2-2 Furnace Performance

Furnace performance sheets should specify the input capacity, output capacity, temperature rise, and AFUE. These sheets also may specify the steady-state efficiency (as a percentage of input capacity), the electrical power requirements, and the amount of cooling capacity that can be added to the furnace.

The performance data package should include a table that correlates the air flow that can be delivered by the blower with the resistance that will be produced by the air distribution system. These tables usually have a set of footnotes that identify the air-side components that were in place when the blower test was performed and the components that were not in place during the test. This information is important because a pressure loss is associated with every component that is installed in the air circulation path. The air-side components that are commonly associated with a furnace include:

- Standard filter
- Accessory filters (mechanical or electronic)
- Heat exchanger
- Refrigerant coil (dry)
- Refrigerant coil (wet)

Also look for "features" and "accessories" sheets that provide information on the fuel-train controls, operating controls, safety controls, thermostat options, and filter options. Some examples of furnace performance data sheets are included in this appendix.

A2-3 Cooling Equipment Performance

Cooling equipment performance tables should provide information on total capacity, sensible capacity, and latent capacity. In this regard, multiple sets of capacity data should be correlated with a range of heat sink temperatures (outdoor air temperature or entering water temperature), a matrix of entering conditions (dry-bulb and wet-bulb temperature of air that is just upstream from the evaporator), and a range of CFM values. These tables also should include power draw information.

Note that cooling performance tables usually have footnotes that pertain to operating temperatures, power requirements, and capacity adjustments. Some of the subjects that are commonly addressed in these footnotes include:

- Adjustment for entering dry-bulb temperature
- Adjustment for entering wet-bulb temperature
- Adjustment for altitude
- Adjustment for fan heat
- Equipment associated with power draw values
- Equipment not associated with power draw values

Also look for features and accessories sheets that provide information on the operating controls, safety controls, thermostat options, and filter options. Some examples of cooling performance data sheets are included in this appendix.

A2-4 Heat Pump Heating Performance

Heating performance tables should provide information on heating capacity and power draw for an appropriate range of heat-source temperatures (outdoor air temperature or entering water temperature). The performance data also should account for the effect of the defrost cycle (air-to-air equipment), for variations in air-side and water-side flow rates, and for variations in air-side temperatures.

Always check the footnotes that accompany the performance tables for information about the range of application of the tabular data or for procedures that can be used to modify the tabulated data. Some of the subjects commonly addressed in the footnotes include:

- Defrost penalty accounted for
- Defrost penalty not accounted for
- Correction for variations in the blower CFM
- Correction for variations in the entering air temperature
- Adjustment for fan heat
- Adjustment for altitude
- Comments about electrical resistance heat
- Equipment associated with power draw values
- Equipment not associated with power draw values

Also look for features and accessories sheets that provide information on resistance heating coils, defrost control options, operating controls, safety controls, thermostat options, thermostat sub-base options, and filter options. Some examples of heating performance data sheets are included in this appendix.

A2-5 Blower Performance

Heat pump and cooling equipment data packages should include tables that correlate the air flow that can be delivered by the blower with the resistance that will be produced by the air distribution system. These tables usually have a set of footnotes that identify the air-side components that were in place when the blower test was performed and the components that were not in place during the test. This information is important because a pressure loss is associated with every component that is installed in the air circulation path. The air-side components normally associated with the indoor air handling package include:

- Refrigerant coil (dry)
- Refrigerant coil (wet)
- Standard filter
- Accessory filters (media)
- Accessory filters (electronic)

- Electric resistance heating coils (heat pumps)
- Humidifiers

A2-6 Component Pressure Drops

Pressure drop information is required for any component that is not accounted for in the blower table. This information is found in supplemental tables that are packaged with the equipment performance data. Components that may not be accounted for in a blower table include:

- Refrigerant coils installed in furnaces
- Coils used with three-piece equipment packages
- Wet coil allowance
- Resistance heaters installed in heat pump systems
- Standard filter
- Electronic or high-efficiency filters
- Humidifiers

Furnace Data

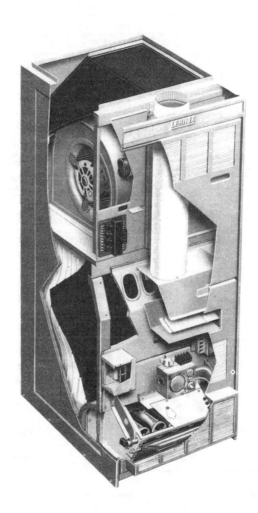

SPECIFICATIONS

Model No.		G17RQ2/3-50	G17RQ3-75	G17RQ3/4-100	G17RQ5-100	G17RQ4/5-125
Input Btuh		50,000	75,000	100,000	100,000	125,000
Output Btuh		40,000	58,000	79,000	77,000	99,000
*A.F.U.E.		78.0%	78.0%	78.0%	78.0%	78.0%
Flue size connection (in. diameter)		4 round	4 round	5 oval	5 oval	6 oval
Temperature rise range (°F)		20-50	50-80	50-80	40-70	50-80
High static certified by A.G.A. (in wg.)		0.50	0.50	0.50	0.50	0.50
Gas Piping Size	Natural	1/2	1/2	1/2	1/2	1/2
I.P.S. (in.)	**LPG	1/2	1/2	1/2	1/2	1/2
Blower wheel nominal diameter x width (in.)		10 x 8	10 x 8	11 x 9	12 x 12	12 x 12
Blower motor hp		1/3	1/3	1/2	3/4	3/4
Number and size of filters (in.)		(2) 20 x 10 x 1	(2) 20 x 10 x 1	(2) 20 x 12 x 1	(2) 20 x 14 x 1	(2) 20 x 14 x 1
Tons of cooling that can be added		2, 2-1/2 or 3	2, 2-1/2 or 3	3, 3-1/2 or 4	4 or 5	4 or 5
Shipping weight (lbs.)		175	191	243	246	276
Number of packages in shipment		1	1	1	1	1
Electrical characteristics		120 volts — 60 hertz — 1 phase (less than 12 amps) All models				
**LPG kit (Optional)		LB-62384DD (All models)				
Continuous Low Speed Blower Kit (Optional)		LB-63646A (All models)				
Down-Flo Additive	Part No.	LB-80639BA	LB-80639BA	LB-80639BB	LB-80639BC	LB-80639BC
Base (Optional)	Shipping weight (lbs.)	6	6	6	8	8

*Annual Fuel Utilization Efficiency based on D.O.E. test procedures and according to F.T.C. labeling requirements. Isolated combustion system rating for non-weatherized furnaces.
**LPG kit must be ordered extra for field changeover.

A.G.A. INSTALLATION CLEARANCES

Sides	1 inch
Rear	1 inch
Top	1 inch
**Front	**6 inches
***Floor	***Combustible
•Flue	•1 inch
*Flue	*6 inches

•Type "B" vent clearances as listed by U.L.
*This is clearance to all flue pipes except type "B".
**Measured from the draft hood relief opening.
NOTE—Flue sizing and air for combustion and ventilation must conform to the methods outlined in American National Standard (ANSI-Z223.1) National Fuel Gas Code.
***Clearance for installation on combustible floor if optional additive base is installed between the furnace and the combustible floor. Not required in add-on cooling coil applications if installed in accordance with local codes or National Fuel Gas Code ANSI-Z223.1.

HIGH ALTITUDE DERATE

A.G.A. certified units must be derated when installed at an elevation of more than 2000 feet above sea level. If unit is installed at an altitude higher than 2000 feet, the unit must be derated 4% for every 1000 feet above sea level. Thus, at an altitude of 4000 feet, the unit would require a derate of 16%.

NOTE — This is the only permissible derate for the units.

BLOWER DATA

G17RQ2/3-50 BLOWER PERFORMANCE

External Static Pressure (in. wg)	Air Volume (cfm) @ Various Speeds			
	High	Med-High	Med-Low	Low
0	1660	1420	1155	965
.05	1630	1400	1140	965
.10	1600	1380	1125	970
.15	1560	1350	1115	965
.20	1520	1320	1100	955
.25	1460	1295	1080	945
.30	1400	1265	1055	935
.40	1290	1170	1000	890
.50	1205	1065	885	805
.60	1045	905	800	705
.70	860	780	675	605
.80	710	620	550	475

NOTE — All cfm data is measured external to unit with air filter in place.

G17RQ3-75 BLOWER PERFORMANCE

External Static Pressure (in. wg)	Air Volume (cfm) @ Various Speeds			
	High	Med-High	Med-Low	Low
0	1340	1250	1065	880
.05	1310	1220	1050	875
.10	1280	1195	1030	865
.15	1250	1165	1010	855
.20	1220	1135	990	840
.25	1195	1105	965	825
.30	1165	1075	940	805
.40	1070	1000	880	755
.50	990	915	815	700
.60	895	835	750	645
.70	815	760	665	565
.80	735	660	575	470

NOTE — All cfm data is measured external to unit with air filter in place.

G17RQ3/4-100 BLOWER PERFORMANCE

External Static Pressure (in. wg)	Air Volume (cfm) @ Various Speeds			
	High	Med-High	Med-Low	Low
0	1805	1485	1300	1005
.05	1780	1470	1290	995
.10	1755	1455	1280	985
.15	1720	1440	1265	965
.20	1685	1425	1245	945
.25	1650	1400	1215	925
.30	1615	1370	1190	910
.40	1515	1315	1140	870
.50	1430	1250	1085	825
.60	1330	1155	1025	765
.70	1225	1065	920	700
.80	1100	935	840	630

NOTE — All cfm data is measured external to unit with air filter in place.

G17RQ5-100 BLOWER PERFORMANCE

External Static Pressure (in. wg)	Air Volume (cfm) @ Various Speeds				
	High	Med-High	Medium	Med-Low	Low
0	2510	2335	2115	1910	1690
.05	2470	2300	2085	1880	1665
.10	2430	2260	2055	1855	1645
.15	2390	2230	2020	1820	1610
.20	2350	2200	1980	1790	1580
.25	2310	2165	1940	1740	1535
.30	2270	2125	1905	1690	1490
.40	2175	2030	1825	1625	1400
.50	2080	1940	1730	1535	1330
.60	1990	1835	1640	1450	1245
.70	1880	1735	1545	1350	1155
.80	1785	1635	1470	1280	1075

NOTE — All cfm data is measured external to unit with air filter in place.

G17RQ4/5-125 BLOWER PERFORMANCE

External Static Pressure (in. wg)	Air Volume (cfm) @ Various Speeds				
	High	Med-High	Medium	Med-Low	Low
0	2415	2215	1980	1770	1575
.05	2390	2180	1960	1750	1555
.10	2360	2150	1945	1730	1535
.15	2325	2115	1915	1710	1515
.20	2290	2085	1890	1690	1495
.25	2245	2055	1860	1665	1485
.30	2195	2025	1825	1640	1475
.40	2135	1970	1775	1595	1425
.50	2050	1910	1720	1540	1385
.60	1965	1835	1685	1485	1325
.70	1905	1765	1605	1425	1260
.80	1835	1695	1545	1355	1195

NOTE- All cfm data is measured external to unit with air filter in place.

FEATURES

Aluminized Steel Burners — Each burner has four rows of practically continuous ports which result in quiet and clean combustion. A crossover igniter of actual burner ports, perpendicular to the main burner, carries a positive flame from burner to burner to achieve quiet and sure ignition.

Electronic Pilot Ignition — Solid-state electronic spark igniter provides positive ignition of pilot burner on each operating cycle. Pilot gas is ignited and burns during each running cycle (intermittent pilot) of the furnace. Main burners and pilot gas are extinguished during the off cycle. This system permits main gas valve to open only when the pilot burner is proven to be lit. Should a loss of flame occur, the main valve closes, shutting down the unit. Pilot is a fully automatic operation on demand for heat.

Automatic Gas Control — Silent operating gas controls provide 100% safety shut off. 24 volt redundant combination gas control valve combines automatic safety pilot, manual shut off knob (On-Off), pilot filtration, automatic electric valve (dual) and gas pressure regulation into a compact combination control. Dual valve design provides double assurance of 100% close off of gas to the pilot and main burners on each off cycle.

Rugged Cabinet — Constructed of heavy gauge cold rolled steel. Cabinet is subject to five station metal wash process resulting in a perfect bonding surface for a paint finish of baked-on enamel. The paint solution and metal are given opposite electrical charges resulting in positive adhesion and even coverage of the paint to the metal surfaces. Cabinet surface temperatures are low due to interior metal liners on each side of cabinet and foil faced fiberglass insulation on vestibule panel, side panels and on back panel. Complete service access is accomplished by removing furnace and blower compartment doors and access panels. Safety interlock switch located on the blower vestibule panel automatically shuts off power to the unit when filter access door is removed. Gas piping and electrical inlet knockouts are provided in both sides of the cabinet. Return air opening is flanged for ease of duct connection. Supply air opening matches the supply air opening in add-on Lennox down-flo evaporator coils.

Combustion Air Damper — Damper is factory installed in the aluminized steel burner box extension of the heat exchanger. Energy saving damper closes off combustion air flow through the heat exchanger during burner off cycle to prevent loss of heated air up the flue. Heavy gauge aluminized steel damper is gasketed for tight seal and rotates smoothly in nylon bearings. Equipped with a heavy duty synchronous spring return damper motor. Removable top on burner box allows access into the burner area for servicing and field conversion to LPG. Damper proving switch confirms that damper is open before allowing main gas valve to open. An observation port with cover is furnished on burner box for flame viewing

Flame Rollout Switch — Manual reset switch is furnished as standard and is factory installed on the burner box. Switch prevents unit operation in the event combustion products passage through the flueway is reduced or blocked.

Blocked Vent Safety Shutoff Sensor — Manual reset temperature sensor prevents unit operation in case of flue blockage and meets ANSI requirements. Sensor is furnished as standard and is factory installed on the draft hood.

Limit Controls — Factory installed and accurately located limit controls (dual) have fixed temperature settings and are located in furnace and blower sections. Protects unit in case of abnormal operating conditions.

BCC2-2 Blower Control Center — Furnished and factory installed on blower vestibule panel. Solid-state board contains all necessary controls and relays to operate furnace. Fan control consists of adjustable blower timed-off delay (90 to 330 seconds) and fixed blower timed-on delay (45 seconds). For air-conditioning applications, blower is automatically energized on thermostat demand for cooling. Provisions have been made for additional wiring connections required for power humidifiers and electronic air cleaners. Also included is a low voltage terminal strip for thermostat connections.

Wiring Junction Box — Power supply connections are made at the wiring junction box which is located on the furnace vestibule panel. Box contains control transformer.

Transformer — 24 volt control transformer is furnished as standard equipment and is factory installed in wiring junction box.

Powerful Blowers — Units are equipped with quiet multi-speed direct drive blowers. Each blower assembly is statically and dynamically balanced. Multiple-speed leadless motor is resiliently mounted. A choice of blower speeds is available on each blower. See blower performance tables.

Cleanable Air Filters — Washable or vacuum cleanable frame type filters are furnished as standard. Polyurethane media is coated with oil for increased efficiency. Factory installed filter rack is furnished for easy filter replacement.

OPTIONAL EQUIPMENT (Must Be Ordered Extra)

Continuous Low Speed Blower Kit (Optional) — Field installed kit (LB-63646A) is available to provide continuous low speed blower operation. Kit includes switch and all necessary wiring. Kit is not furnished and must be ordered extra.

Down-flo Additive Base (Optional) — Additive base is required for heating only units installed on combustible floors. Base is not furnished and must be ordered extra for field installation. See specifications table. Not required in add-on cooling applications.

LPG Conversion Kit (Optional) — For LPG models a conversion kit is required for field changeover from natural gas. Kit is not furnished and must be ordered extra. See specifications table for order number.

Thermostat (Optional) — Heating thermostat is not furnished and must be ordered extra. See Accessories Section, Page 13 and Lennox Price Book. For all-season applications, heating and cooling thermostat is available with the condensing unit.

Furnace Cooling Coil Data

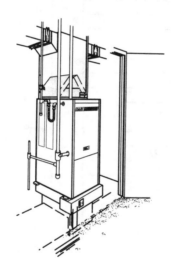

AIR RESISTANCE

Model No.	Air Volume (cfm)	Total Resistance (in. wg.)
C16-18FF	500	.08
	600	.10
	700	.13
	750	.15
C16-21FF	600	.08
	800	.11
	1000	.15
	1200	.19
C16-28FF	800	.08
	1000	.12
	1200	.17
	1400	.23
C16-28WFF	800	.07
	1000	.11
	1200	.16
	1400	.19
	1600	.25
C16-31FF	800	.14
	1000	.17
	1200	.22
	1400	.28
C16-31WFF	800	.12
	1000	.15
	1200	.20
	1400	.24
	1600	.30
C16-41FF	1000	.13
	1200	.19
	1400	.26
	1600	.34
C16-41WFF	1000	.11
	1200	.15
	1400	.21
	1600	.27
	1800	.35
C16-46FF	1200	.09
	1400	.12
	1600	.18
	1800	.25
C16-46WFF	1200	.08
	1400	.11
	1600	.16
	1800	.22
	2000	.27
C16-51FF	1200	.13
	1400	.14
	1600	.16
	1800	.18
	2000	.19
C16-65	1400	.13
	1600	.14
	1800	.16
	2000	.18
	2200	.20
	2400	.22
	2600	.24

EXPANSION VALVE KIT SELECTOR

Evaporator Model No.	Condensing Unit	*Expansion Valve Kit
C16-18FF	HS18-141	LB-25778CH
	HS18-211	LB-25778CG
C16-21FF	HS16-211V	LB-53081CF
	HS16-261V, HS19-261V	LB-53081CD
	HS18-141	LB-25778CH
	HS18-211, HS18-261	LB-25778CG
C16-28FF C16-28WFF C16-31FF C16-31WFF	HS14-411-413V	LB-53081CA
	HS16-211V	LB-53081CF
	HS16-261V, HS16-311V, HS19-261V, HS19-311V	LB-53081CD
	HS16-411V, HS19-411V	LB-53081CB
	HS18-211, HS18-261	LB-25778CG
	HS18-311	LB-25778CE
	HS18-411-413	LB-25778CF
C16-41FF C16-41WFF	HS14-411-413V	LB-53081CA
	HS16-261V, HS16-311V, HS19-261V, HS19-311V	LB-53081CD
	HS16-411V, HS16-461V, HS19-411V, HS19-461V	LB-53081CB
	HS18-311	LB-25778CE
	HS18-411-413, HS18-461-463	LB-25778CF
C16-46FF C1-46WFF	HS14-411-413V	LB-53081CA
	HS14-511-513V, HS16-411V, HS16-461V, HS19-411V, HS19-461V	LB-53081CB
	HS16-511V, HS19-511V	LB-53081CC
	HS18-411-413, HS18-461-463	LB-25778CF
	HS18-511-513	LB-25778CC
C16-51FF	HS14-511-513V, HS16-411V, HS16-461V, HS19-411V, HS19-461V	LB-53081CB
	HS14-651-653V, HS16-511V, HS19-511V	LB-53081CC
	HS16-651V, HS19-651V	LB-53081CE
	HS18-461-463	LB-25778CF
	HS18-511-513	LB-25778CC
	HS18-651-653	LB-25778CD
C16-65	HS14-511-513V	LB-53081CB
	HS14-651-653V, HS16-511V, HS19-511V	LB-53081CC
	HS16-651V, HS19-651V	LB-53081CE
	HS18-511-513	LB-25778CC
	HS17-813V, HS18-651-653	LB-25778CD

*Must be ordered extra for field installation.

Air-to-Air Cooling Equipment

RATINGS

NOTE — To determine sensible capacity, leaving wet bulb and dry bulb temperatures not shown in the tables, see Miscellaneous Engineering Data, page 9.

HS14-511V-513V WITH C16-46FF(FC) OR C16-46WFF(FC) EVAPORATOR UNIT
(Low Speed Compressor Operation)

Enter. Wet Bulb (°F)	Total Air Vol. (cfm)	Outdoor Air Temperature Entering Condenser Coil (°F)																
		75					85					95					105				
		Total Cool Cap. (Btuh)	Comp. Motor Watts Input	Sensible To Total Ratio (S/T) Dry Bulb (°F)			Total Cool Cap. (Btuh)	Comp. Motor Watts Input	Sensible To Total Ratio (S/T) Dry Bulb (°F)			Total Cool Cap. (Btuh)	Comp. Motor Watts Input	Sensible To Total Ratio (S/T) Dry Bulb (°F)			Total Cool Cap. (Btuh)	Comp. Motor Watts Input	Sensible To Total Ratio (S/T) Dry Bulb (°F)		
				76	80	84			76	80	84			76	80	84			76	80	84
63	1200	31,700	1770	.83	.93	1.00	30,300	1890	.86	.93	1.00	28,800	2040	.89	.93	1.00	27,100	2220	.93	1.00	1.00
	1600	34,400	1780	.93	1.00	1.00	32,700	1900	.93	1.00	1.00	30,800	2060	.93	1.00	1.00	28,800	2260	1.00	1.00	1.00
	2000	36,300	1780	1.00	1.00	1.00	34,300	1910	1.00	1.00	1.00	32,200	2080	1.00	1.00	1.00	29,900	2280	1.00	1.00	1.00
67	1200	33,800	1780	.63	.77	.90	31,900	1900	.65	.79	.93	29,800	2050	.67	.82	.93	27,700	2240	.69	.86	.93
	1600	35,200	1780	.70	.87	.93	33,200	1900	.72	.90	1.00	30,900	2060	.75	.93	1.00	28,800	2260	.79	.93	1.00
	2000	36,300	1780	.77	.93	1.00	34,400	1910	.79	.93	1.00	32,200	2080	.83	1.00	1.00	30,000	2290	.87	1.00	1.00
71	1200	36,600	1780	.45	.58	.71	34,400	1910	.46	.60	.73	32,000	2070	.47	.62	.76	29,500	2280	.49	.64	.80
	1600	37,800	1790	.49	.64	.80	35,400	1920	.50	.67	.83	32,800	2090	.51	.70	.87	30,200	2290	.53	.73	.92
	2000	38,600	1790	.52	.71	.89	36,100	1920	.53	.74	.93	33,400	2090	.55	.77	.93	30,700	2300	.58	.81	.93

NOTE — All values are gross capacities and do not include evaporator coil blower motor heat deduction.

HS14-511V-513V WITH C16-46FF(FC) OR C16-46WFF(FC) EVAPORATOR UNIT
(High Speed Compressor Operation)

Enter. Wet Bulb (°F)	Total Air Vol. (cfm)	Outdoor Air Temperature Entering Condenser Coil (°F)																
		85					95					105					115				
		Total Cool Cap. (Btuh)	Comp. Motor Watts Input	Sensible To Total Ratio (S/T) Dry Bulb (°F)			Total Cool Cap. (Btuh)	Comp. Motor Watts Input	Sensible To Total Ratio (S/T) Dry Bulb (°F)			Total Cool Cap. (Btuh)	Comp. Motor Watts Input	Sensible To Total Ratio (S/T) Dry Bulb (°F)			Total Cool Cap. (Btuh)	Comp. Motor Watts Input	Sensible To Total Ratio (S/T) Dry Bulb (°F)		
				76	80	84			76	80	84			76	80	84			76	80	84
63	1200	46,000	3990	.74	.84	.94	44,000	4250	.75	.86	.96	41,900	4520	.76	.88	.98	39,900	4770	.78	.90	1.00
	1600	48,800	4100	.80	.92	1.00	46,500	4370	.82	.95	1.00	44,400	4640	.84	.97	1.00	41,900	4900	.86	1.00	1.00
	2000	50,600	4170	.86	1.00	1.00	48,500	4460	.89	1.00	1.00	46,400	4760	.91	1.00	1.00	44,300	5040	.94	1.00	1.00
67	1200	49,700	4130	.58	.68	.78	47,400	4410	.59	.69	.79	45,100	4690	.60	.71	.81	42,800	4950	.61	.72	.83
	1600	52,200	4230	.62	.74	.86	49,600	4510	.63	.76	.88	47,000	4790	.65	.78	.91	44,600	5060	.66	.80	.93
	2000	53,700	4280	.66	.80	.94	51,100	4570	.68	.82	.96	48,400	4860	.69	.85	.99	45,800	5130	.71	.88	1.00
71	1200	53,600	4280	.45	.54	.63	51,000	4570	.45	.54	.64	48,500	4860	.45	.55	.65	46,000	5140	.46	.56	.67
	1600	56,000	4360	.46	.57	.69	53,200	4660	.47	.58	.70	50,400	4960	.47	.60	.72	47,700	5240	.48	.61	.74
	2000	57,500	4420	.48	.61	.74	54,500	4720	.49	.63	.76	51,600	5020	.50	.64	.79	48,700	5300	.50	.66	.82

NOTE — All values are gross capacities and do not include evaporator coil blower motor heat deduction.

HS14-511V-513V WITH CR16-51FF EVAPORATOR UNIT
(Low Speed Compressor Operation)

Enter. Wet Bulb (°F)	Total Air Vol. (cfm)	Outdoor Air Temperature Entering Condenser Coil (°F)																
		75					85					95					105				
		Total Cool Cap. (Btuh)	Comp. Motor Watts Input	Sensible To Total Ratio (S/T) Dry Bulb (°F)			Total Cool Cap. (Btuh)	Comp. Motor Watts Input	Sensible To Total Ratio (S/T) Dry Bulb (°F)			Total Cool Cap. (Btuh)	Comp. Motor Watts Input	Sensible To Total Ratio (S/T) Dry Bulb (°F)			Total Cool Cap. (Btuh)	Comp. Motor Watts Input	Sensible To Total Ratio (S/T) Dry Bulb (°F)		
				76	80	84			76	80	84			76	80	84			76	80	84
63	1200	32,000	1780	.83	.94	1.00	30,700	1890	.85	.94	1.00	29,100	2040	.88	.94	1.00	27,400	2220	.92	.94	1.00
	1600	34,800	1780	.93	1.00	1.00	33,100	1900	.94	1.00	1.00	31,200	2060	.94	1.00	1.00	29,200	2260	.94	1.00	1.00
	2000	36,800	1780	1.00	1.00	1.00	34,800	1910	1.00	1.00	1.00	32,700	2080	1.00	1.00	1.00	30,400	2290	1.00	1.00	1.00
67	1200	34,400	1780	.63	.76	.89	32,500	1900	.64	.79	.92	30,400	2050	.67	.82	.94	28,200	2240	.69	.85	.94
	1600	35,900	1780	.69	.86	.94	33,800	1910	.72	.89	.94	31,600	2060	.74	.93	1.00	29,200	2260	.78	.94	1.00
	2000	37,500	1780	.76	.97	1.00	35,200	1910	.81	1.00	1.00	32,800	2070	.82	1.00	1.00	30,200	2270	.88	1.00	1.00
71	1200	37,400	1790	.45	.58	.70	35,200	1920	.46	.59	.73	32,700	2080	.47	.61	.76	30,100	2280	.48	.64	.79
	1600	38,700	1790	.48	.64	.79	36,200	1920	.50	.66	.82	33,600	2090	.51	.69	.86	30,800	2300	.53	.72	.91
	2000	39,500	1790	.51	.70	.87	36,900	1930	.53	.72	.91	34,100	2100	.55	.76	.94	31,300	2310	.57	.80	.94

NOTE — All values are gross capacities and do not include evaporator coil blower motor heat deduction.

HS14-511V-513V WITH CR16-51FF EVAPORATOR UNIT
(High Speed Compressor Operation)

Enter. Wet Bulb (°F)	Total Air Vol. (cfm)	Outdoor Air Temperature Entering Condenser Coil (°F)																
		85					95					105					115				
		Total Cool Cap. (Btuh)	Comp. Motor Watts Input	Sensible To Total Ratio (S/T) Dry Bulb (°F)			Total Cool Cap. (Btuh)	Comp. Motor Watts Input	Sensible To Total Ratio (S/T) Dry Bulb (°F)			Total Cool Cap. (Btuh)	Comp. Motor Watts Input	Sensible To Total Ratio (S/T) Dry Bulb (°F)			Total Cool Cap. (Btuh)	Comp. Motor Watts Input	Sensible To Total Ratio (S/T) Dry Bulb (°F)		
				76	80	84			76	80	84			76	80	84			76	80	84
63	1200	46,400	3950	.73	.83	.93	44,300	4210	.74	.85	.95	42,200	4470	.75	.86	.97	40,200	4720	.77	.89	1.00
	1600	49,200	4060	.79	.91	1.00	46,900	4330	.80	.93	1.00	44,600	4600	.82	.95	1.00	42,400	4860	.85	.98	1.00
	2000	51,300	4130	.84	.98	1.00	48,600	4410	.87	1.00	1.00	46,500	4700	.89	1.00	1.00	44,400	4980	.92	1.00	1.00
67	1200	50,200	4100	.58	.67	.77	47,900	4370	.58	.68	.78	45,600	4650	.59	.70	.80	43,200	4910	.60	.71	
	1600	52,800	4190	.61	.73	.84	50,200	4470	.62	.74	.86	47,600	4750	.63	.76	.89	45,100	5020	.65	.78	.91
	2000	54,500	4250	.65	.78	.91	51,700	4540	.66	.80	.94	48,900	4820	.68	.83	.97	46,300	5090	.70	.85	1.00
71	1200	54,200	4240	.44	.53	.62	51,600	4540	.45	.54	.63	49,000	4830	.45	.55	.64	46,500	5100	.45	.56	.66
	1600	56,700	4330	.46	.56	.67	53,900	4630	.46	.58	.69	51,100	4920	.47	.59	.71	48,300	5210	.47	.60	.73
	2000	58,300	4390	.47	.60	.72	55,300	4690	.48	.61	.75	52,300	4990	.49	.63	.77	49,400	5270	.50	.65	.79

NOTE — All values are gross capacities and do not include evaporator coil blower motor heat deduction.

ARI RATINGS

Condensing Unit Model No. ★ARI Standard 270 SRN (bels)	*ARI Standard 210/240 Ratings				Evaporator Unit			☆Expansion Valve Kit
	SEER (Btuh/Watt)	EER (Btuh/Watt)	Cooling Capacity (Btuh)	Total Unit Watts	Up-Flo	Down-Flo	Horizontal	
HS14-411V HS14-413V (7.6)	11.85	9.20	32,800	3556	----	CR16-31FF	----	LB-53081CA
	12.05	9.40	33,800	3593	C16-28FF(FC), C16-28WFF(FC) C16-31FF(FC), C16-31WFF(FC)	----	----	
	12.15	9.40	33,800	3593	----	----	CH16-31FF	
	12.70	9.80	35,200	3585	**CB18-31	----	**CBS18-31	
	12.45	9.80	35,800	3652	----	CR16-41FF	----	
	12.45	9.80	35,800	3655	C16-41FF(FC) C16-41WFF(FC)	----	----	
	12.85	9.95	36,000	3610	**CB18-41	----	**CBS18-41	
	12.70	9.90	36,400	3675	----	----	CH16-41FF	
	12.45	9.95	36,600	3680	----	CR16-51FF	----	
	12.45	9.95	36,600	3678	C16-46FF(FC) C16-46WFF(FC)	----	----	
	12.80	10.00	37,000	3705	**CB18-51	----	**CBS18-51	
	12.75	10.05	37,400	3724	----	----	CH16-51FF	
	13.30	10.40	38,000	3698	C14-41FF(FC)	----	----	
	12.90	10.50	38,500	3650	**CB19-31	**CB19-31	**CBH19-31	
	12.95	10.50	39,000	3714	**CB19-41	**CB19-41	**CBH19-41	
	15.00	10.80	39,500	3665	**CB21-41	**CB21-41	CBH21-41	
	12.50	10.70	40,500	3793	**CB19-51	**CB19-51	**CBH19-51	
	15.50	10.80	43,000	3980	**CB21-51	**CB21-51	**CBH21-51	
HS14-511V HS14-513V (7.6)	11.35	9.10	46,500	5110	C16-46FF(FC) C16-46WFF(FC)	----	----	LB-53081CB
	11.65	9.25	47,000	5074	----	CR16-51FF	----	
	11.65	9.25	47,500	5140	----	----	CH16-51FF	
	11.85	9.30	47,500	5104	**CB18-51	----	**CBS18-51	
	11.55	9.40	48,000	5119	C16-51FF(FC)	----	----	
	11.45	9.15	48,500	5307	**CB18-65	----	**CBS18-65	
	11.90	9.50	49,000	5159	C16-65(FC)	CR16-65	----	
	12.55	9.65	49,000	5080	C14-41FF(FC)	----	----	
	12.40	9.50	49,500	5201	C14-65(FC)	----	----	
	12.10	9.55	50,000	5229	----	----	CH16-65V	•Factory Installed
	12.50	10.00	51,000	5100	**CB19-51	**CB19-51	**CBH19-51	LB-53081CB
	12.30	10.00	53,000	5293	**CB19-65	**CB19-65	**CBH19-65	
	13.40	9.70	53,000	5440	**CB21-51	**CB21-51	**CBH21-51	
	13.30	9.70	54,000	5550	**CB21-65	**CB21-65	**CBH21-65	
HS14-651V HS14-653V (7.8)	11.10	8.15	55,000	6774	----	----	CH16-51FF	LB-53081CC
	11.40	8.00	55,500	6973	----	CR16-51FF	----	
	11.35	8.25	56,000	6819	C14-41FF(FC)	----	----	
	10.80	8.10	56,500	6993	C16-51FF(FC)	----	----	
	10.80	8.15	58,000	7129	**CB18-51	----	**CBS18-51	
	10.80	8.20	58,000	7079	----	CR16-65FF	----	
	10.90	8.30	59,000	7115	C16-65(FC)	----	----	
	10.85	8.20	59,500	7307	**CB18-65	----	**CBS18-65	
	11.40	8.40	59,500	7072	**CB19-51	**CB19-51	**CBH19-51	
	11.40	8.55	61,500	7236	----	----	CH16-65V	•Factory Installed
	11.65	8.60	61,500	7187	C14-65(FC)	----	----	LB-53081CC
	12.75	9.10	63,000	6935	**CB21-51	**CB21-51	**CBH21-51	
	12.60	8.90	63,500	7152	**CB19-65	**CB19-65	**CBH19-65	
	12.65	9.30	64,500	6970	**CB21-65	**CB21-65	**CBH21-65	

NOTE — Shaded area denotes most popular evaporator coil.
★ Sound Rating Number in accordance with ARI Standard 270.
* Rated in accordance with ARI Standard 210/240 and DOE; 95°F outdoor air temperature, 80°F db/67°F wb entering evaporator air with 25 ft. of connecting refrigerant lines.
** Denotes blower powered evaporator.
☆ Kit is optional and must be ordered extra for field installation.
• Furnished as standard with coil.

Durable Steel Cabinet — Heavy gauge galvanized steel cabinet is subject to a five station metal wash process. This preparation process results in a perfect bonding surface for the finish coat of baked-on outdoor enamel. The attractive enamel finish gives the cabinet long lasting protection from the weather. Drainage holes are furnished in base section for moisture removal. Heavy duty steel base channels raise the unit off of the mounting surface away from damaging moisture. A non-corrosive PVC coated steel wire condenser coil guard is furnished on all models.

Accessible Control Box — Large size and conveniently located in the compressor and controls compartment for easy access. All controls are pre-wired at the factory.

Compressor and Controls Compartment — Separate compressor and controls compartment protects all components from weather conditions and keeps sound transmission at a minimum. Large removable access panel is lined with thick fiberglass insulation and provides complete service access.

Efficient Two-Speed Compressor — The two-speed compressor is designed for superior operating efficiency at minimum cost. Two speed operation gives staging control to fit varying cooling load requirements, extends service life of the compressor and provides operation economy during periods of reduced loads. During part load conditions the compressor operates in the low speed mode. Operates at 1750 rpm at low speed and 3500 rpm at high speed. Reliable compressor is hermetically sealed with built-in protection from excessive current and temperatures. Suction cooled and overload protected. Equipped with solid-state motor protection, vertical crankshaft, ringed valves and pistons, tuned discharge muffler, two stage oil pump and positive venting of lube system. Crankcase heater assures proper compressor lubrication. The entire running gear assembly is resiliently suspended internally. In addition, the compressor is installed in the unit on resilient rubber mounts assuring low sound and vibration free operation.

Quiet Condenser Fan — Efficient direct drive fan moves large volumes of air uniformly through the entire condenser coil resulting in high refrigerant cooling capacity. Vertical discharge of air minimizes operating sounds and eliminates hot air damage to lawn and shrubs. Fan motor has permanently lubricated ball bearings, is inherently protected and totally enclosed for maximum protection from weather, dust and corrosion. A rain shield on the motor provides additional protection from moisture. Fan service access is accomplished by removal of fan guard. Corrosion resistant PVC coated steel wire fan guard is furnished as standard.

Copper Tube/Enhanced Fin Coil — Lennox designed and fabricated coil is constructed of precisely spaced ripple-edged aluminum fins machine fitted to seamless copper tubes in a wrap around "U" shaped configuration providing extra large surface area with low air resistance. Lanced fins provide maximum exposure of fin surface to air stream resulting in excellent heat transfer. In addition, fins are equipped with collars that grip the tubing for maximum contact area. Precise circuiting provides uniform refrigerant distribution. Flared shoulder tubing connections and silver soldering provide tight, leakproof joints. Long life copper tubing is corrosion-resistant and easy to field service. Coil is thoroughly factory tested under high pressure to insure leakproof construction. Entire coil is accessible for cleaning.

Hi-Capacity Drier — Furnished and factory installed. Drier traps any moisture or dirt that could contaminate the refrigerant system.

High Pressure Switch — Shuts off the unit if abnormal operat conditions cause the discharge pressure to rise above settin Switch protects the compressor from excessive condens pressure. Manual reset.

Low Pressure Switch — Shuts off unit if suction pressure fa below setting. Provides loss of charge and freeze-up protecti Automatic reset.

Start Controls — Furnished and factory installed. Provides as tance for compressor start under loaded conditions or in the ev of low voltage.

Lennox TSC-2 Timed Start Control Module — Furnished a factory installed. Prevents compressor short-cycling and also allo time for suction and discharge pressure to equalize, permitting compressor to start in an unloaded condition. Module also p vides a time delay between compressor shutoff and start-up a between speed changes.

Refrigerant Line Connections, Electrical Inlets and Serv Valves — Suction and liquid lines are stubbed inside the cabi and are made with sweat connections. Brass service valves p vent corrosion and provide access to refrigerant system. Suct and liquid line service valves and gauge ports are located ins the cabinet. A thermometer well is located in the liquid line to ch the refrigerant charge. Refrigerant line and field wiring inlets all located in one central area of the cabinet. See dimens drawing for location.

Thermostat (Optional) — Thermostat is not furnished with unit and must be ordered extra. See Accessories Section 13 and Lennox Price Book.

HS14-411V-413V Refrigerant Line Kits (Optional) — Ref erant lines are available for the HS14-411V-413V models only must be ordered extra. See refrigerant line kit table. The refrige lines (suction & liquid) are shipped refrigeration clean. Lines cleaned, dried, pressurized and sealed at the factory. Suction is fully insulated. Lines are furnished with a flare fitting (evapora unit connection) on one end and less any fitting (stubbed) on opposite end for connection to the condensing unit.

Indoor Blower Speed Relay Kit (Optional) — Relay kit (72G provides optimum humidity control conditions by automatic reducing indoor blower speed during continuous fan or low sp compressor operation. Kit must be ordered extra and field install

Expansion Valve Kits (Optional) — Must be ordered extra field installed on most evaporator units. See ARI Ratings tal

Low Ambient Kit (Optional) — Condensing units will ope satisfactorily down to 45°F outdoor air temperature without additional controls. For cases where operation of the unit is quired below 45°F a Low Ambient Control Kit (LB-57113BC) be added in the field, enabling it to operate properly down to 0

Mounting Base (Optional) — Rugged MB1-32 mounting b (83C83) provides permanent foundation for condensing uni density polyethylene structural material is lightweight, s sound absorbing and will withstand the rigors of the sun, h cold, moisture, oil and refrigerant. Will not mildew or rot. Can shipped singly or in packages of 6 to a carton. Shipping wei 15 lbs. each. Dimensions 32" x 34" x 3".

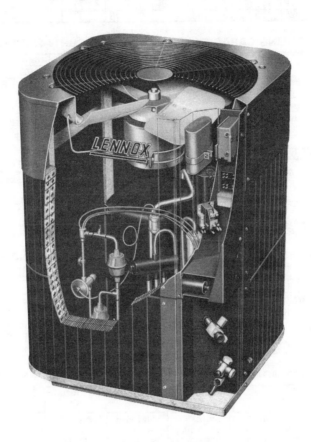

Air-to-Air Heat Pump Equipment

RATINGS

NOTE — To determine sensible capacity, leaving wet bulb and dry bulb temperatures not shown in the tables, see Miscellaneous Engineering Data, page 9.

HP20-411 COOLING CAPACITY WITH CB18-41 OR CBS18-41 INDOOR COIL UNIT

Enter. Wet Bulb (°F)	Total Air Vol. (cfm)	Outdoor Air Temperature Entering Outdoor Coil (°F)															
		85					95					105					115				
		Total Cool Cap. (Btuh)	Comp. Motor Watts Input	Sensible To Total Ratio (S/T) Dry Bulb (°F)			Total Cool Cap. (Btuh)	Comp. Motor Watts Input	Sensible To Total Ratio (S/T) Dry Bulb (°F)			Total Cool Cap. (Btuh)	Comp. Motor Watts Input	Sensible To Total Ratio (S/T) Dry Bulb (°F)			Total Cool Cap. (Btuh)	Comp. Motor Watts Input	Sensible To Total Ratio (S/T) Dry Bulb (°F)		
				75	80	85			75	80	85			75	80	85			75	80	85
63	1050	33,000	2560	.75	.89	1.00	31,800	2820	.76	.91	1.00	30,500	3150	.78	.93	1.00	29,200	3560	.79	.95	1.00
	1200	33,900	2580	.78	.93	1.00	32,600	2850	.79	.94	1.00	31,300	3180	.81	.97	1.00	29,800	3580	.83	.98	1.00
	1350	34,700	2590	.81	.96	1.00	33,200	2860	.82	.98	1.00	31,900	3200	.84	.99	1.00	30,500	3610	.86	1.00	1.00
67	1050	34,900	2600	.59	.73	.86	33,500	2870	.60	.74	.87	32,200	3200	.61	.75	.89	30,700	3620	.62	.77	.91
	1200	35,700	2610	.61	.76	.90	34,400	2890	.62	.77	.91	32,900	3230	.63	.79	.93	31,400	3640	.64	.81	.96
	1350	36,400	2630	.63	.78	.93	35,000	2910	.64	.80	.95	33,500	3250	.65	.82	.97	32,000	3660	.66	.84	.99
71	1050	36,600	2640	.45	.58	.70	35,400	2920	.45	.59	.72	33,900	3260	.46	.60	.73	32,400	3670	.46	.61	.75
	1200	37,600	2660	.46	.60	.73	36,200	2940	.46	.61	.75	34,700	3280	.46	.62	.76	33,100	3700	.47	.63	.78
	1350	38,300	2680	.46	.62	.76	36,900	2960	.47	.63	.78	35,300	3300	.47	.64	.80	33,700	3720	.48	.65	.82

NOTE — All values are gross capacities and do not include indoor coil blower motor heat deduction.

HP20-411 COOLING CAPACITY WITH CB19-41 OR CBH19-41 INDOOR COIL UNIT

Enter. Wet Bulb (°F)	Total Air Vol. (cfm)	Outdoor Air Temperature Entering Outdoor Coil (°F)															
		85					95					105					115				
		Total Cool Cap. (Btuh)	Comp. Motor Watts Input	Sensible To Total Ratio (S/T) Dry Bulb (°F)			Total Cool Cap. (Btuh)	Comp. Motor Watts Input	Sensible To Total Ratio (S/T) Dry Bulb (°F)			Total Cool Cap. (Btuh)	Comp. Motor Watts Input	Sensible To Total Ratio (S/T) Dry Bulb (°F)			Total Cool Cap. (Btuh)	Comp. Motor Watts Input	Sensible To Total Ratio (S/T) Dry Bulb (°F)		
				75	80	85			75	80	85			75	80	85			75	80	85
63	1050	33,600	2560	.77	.91	1.00	32,200	2830	.78	.93	1.00	30,800	3160	.80	.95	1.00	29,500	3580	.81	.97	1.00
	1200	34,300	2570	.80	.95	1.00	33,100	2850	.82	.97	1.00	31,800	3190	.83	.98	1.00	30,500	3610	.85	1.00	1.00
	1350	35,300	2590	.83	.98	1.00	34,000	2870	.85	1.00	1.00	32,600	3220	.87	1.00	1.00	31,300	3640	.89	1.00	1.00
67	1050	35,300	2590	.60	.74	.88	34,000	2870	.61	.76	.90	32,600	3210	.62	.77	.92	31,000	3630	.63	.79	.94
	1200	36,100	2610	.62	.78	.92	34,800	2890	.63	.79	.94	33,300	3230	.64	.81	.96	31,700	3660	.65	.83	.98
	1350	36,800	2630	.64	.81	.96	35,400	2910	.65	.83	.97	33,900	3250	.66	.85	.99	32,200	3670	.68	.87	1.00
71	1050	37,100	2640	.44	.59	.72	35,800	2920	.45	.59	.73	34,300	3260	.45	.60	.75	32,800	3690	.46	.61	.77
	1200	38,000	2660	.45	.61	.75	36,600	2940	.46	.62	.77	35,100	3290	.46	.63	.79	33,500	3710	.47	.64	.81
	1350	38,800	2670	.46	.63	.79	37,300	2960	.47	.64	.81	35,700	3310	.47	.65	.83	34,100	3730	.48	.67	.85

NOTE — All values are gross capacities and do not include indoor coil blower motor heat deduction.

HP20-411 HEATING CAPACITY WITH CB18-41 OR CBS18-41 INDOOR COIL UNIT

Indoor Coil Air Volume (cfm) 70°F db	Air Temperature Entering Outdoor Coil (°F)									
	65		45		25		5		−15	
	Total Htg. Cap. (Btuh)	Comp. Mtr. Input (W)	Total Htg. Cap. (Btuh)	Comp. Mtr. Input (W)	Total Htg. Cap. (Btuh)	Comp. Mtr. Input (W)	Total Htg. Cap. (Btuh)	Comp. Mtr. Input (W)	Total Htg. Cap. (Btuh)	Comp. Mtr. Input (W)
1050	39,900	2770	30,800	2560	21,200	2350	15,100	1985	7500	1515
1200	40,400	2670	31,300	2460	21,700	2250	15,600	1885	8000	1415
1350	40,800	2595	31,700	2385	22,100	2175	16,000	1810	8400	1340

NOTE — Heating capacities include the effect of defrost cycles in the temperature range where they occur.

HP20-411 HEATING CAPACITY WITH CB19-41 OR CBH19-41 INDOOR COIL UNIT

Indoor Coil Air Volume (cfm) 70°F db	Air Temperature Entering Outdoor Coil (°F)									
	65		45		25		5		−15	
	Total Htg. Cap. (Btuh)	Comp. Mtr. Input (W)	Total Htg. Cap. (Btuh)	Comp. Mtr. Input (W)	Total Htg. Cap. (Btuh)	Comp. Mtr. Input (W)	Total Htg. Cap. (Btuh)	Comp. Mtr. Input (W)	Total Htg. Cap. (Btuh)	Comp. Mtr. Input (W)
1050	38,900	2650	29,900	2460	20,400	2270	14,600	1920	7300	1460
1200	39,300	2585	30,300	2395	20,800	2205	15,100	1855	7700	1395
1350	39,800	2515	30,800	2325	21,300	2135	15,500	1790	8100	1325

NOTE — Heating capacities include the effect of defrost cycles in the temperature range where they occur.

HP20-411 HEATING PERFORMANCE
at 1200 cfm Indoor Coil Air Volume (CB18-41 or CBS18-41)

*Outdoor Temp. (Degree F)	Compressor Motor Watts Input	Total Output (Btuh)
65	2670	40,400
60	2615	38,300
55	2565	36,200
50	2510	34,100
47	2480	32,800
45	2460	31,300
40	2405	27,400
35	2355	23,600
30	2300	22,700
25	2250	21,700
20	2195	20,800
17	2165	20,200
15	2120	19,400
10	2000	17,500
5	1885	15,600
0	1765	13,700
−5	1650	11,800
−10	1535	9900
−15	1415	8000
−20	1300	6100

HP20-411 HEATING PERFORMANCE
at 1200 cfm Indoor Coil Air Volume (CB19-41 or CBH19-41)

*Outdoor Temp. (Degree F)	Compressor Motor Watts Input	Total Output (Btuh)
65	2585	39,300
60	2535	37,300
55	2490	35,200
50	2440	33,100
47	2415	31,900
45	2395	30,300
40	2345	26,400
35	2295	22,500
30	2250	21,700
25	2205	20,800
20	2160	20,000
17	2130	19,500
15	2085	18,800
10	1970	16,900
5	1855	15,100
0	1740	13,200
−5	1625	11,400
−10	1510	9500
−15	1395	7700
−20	1280	5900

*Outdoor temperature at 70% relative humidity. Indoor temperature at 70°F.   *Outdoor temperature at 70% relative humidity. Indoor temperature at 70°F.

ARI RATINGS

Outdoor Unit Model No. ★ARI Std. 270 SRN (Bels)	†ARI Standard 210/240 Ratings											Indoor Unit	•Check and Expansion Valve Kit
	Cooling Capacity (Btuh)	High Temp. Htg. Cap. (Btuh)	Low Temp. Htg. Cap. (Btuh)	Total Unit Cooling Watts	SEER (Btuh/Watt)	EER (Btuh/Watt)	Total Unit High Temp. Htg. Watts	*HSPF	High Temp. Htg. C.O.P.	Total Unit Low Temp. Htg. Watts	Low Temp. Htg. C.O.P.		
HP20-211 (7.4)	18,000	17,000	10,600	1830	11.00	9.85	1615	6.90	3.08	1470	2.10	CB18-21 CBS18-21	LB-34792BE
HP20-261 (7.6)	23,600	22,600	13,400	2355	10.80	10.00	2135	7.00	3.10	1965	2.00	CB18-26 CBS18-26	LB-34792BE
	23,800	22,600	13,200	2325	11.00	10.20	2040	7.10	3.24	1870	2.06	CB19-26 CBH19-26	
HP20-311 (7.6)	27,800	26,400	15,900	2866	10.90	9.70	2512	6.85	3.08	2284	2.04	CB19-26 CBH19-26	LB-34792BG
	28,200	27,800	17,000	3095	10.05	9.10	2720	6.80	3.00	2430	2.04	CB18-31 CBS18-31	
	28,800	28,000	17,000	3075	10.25	9.35	2685	6.90	3.06	2405	2.08	CB18-41 CBS18-41	
HP20-411 (7.8)	33,000	32,800	20,200	3575	10.25	9.25	3195	6.85	3.00	2890	2.04	CB18-41 CBS18-41	LB-34792BG
	33,400	31,800	19,500	3479	10.90	9.60	3025	6.95	3.08	2747	2.08	CB19-41 CBH19-41	
	34,600	33,000	20,400	3655	10.45	9.45	3150	6.90	3.06	2880	2.08	CB18-51 CBS18-51	
	35,000	32,200	19,500	3425	10.90	10.20	2955	6.80	3.20	2660	2.16	CB21-41 CBH21-41	
HP20-461 (8.2)	38,800	39,500	22,400	4041	10.90	9.60	3640	7.20	3.18	3218	2.04	CB19-41 CBH19-41	LB-34792BG
	39,000	39,000	24,400	3980	11.00	9.80	3790	6.90	3.02	3305	2.18	CB21-41 CBH21-41	
	40,000	40,000	25,200	4270	10.30	9.35	3840	7.15	3.06	3465	2.14	CB18-51 CBS18-51	
	43,000	40,000	23,000	4095	11.60	10.50	3552	7.20	3.30	3209	2.10	CB19-51 CBH19-51	
	43,500	40,000	24,800	4305	11.05	10.10	3690	7.70	3.38	3330	2.18	CB21-51 CBH21-51	

★ Sound Rating Number in accordance with ARI Standard 270.
† Rated in accordance with ARI Standard 210/240 and DOE; with 25 ft. of connecting refrigerant lines.
 Cooling Ratings — 95°F outdoor air temperature and 80°F db/67°F wb entering indoor coil air.
 High Temperature Heating Ratings — 47°F db/43°F wb outdoor air temperature and 70°F db entering indoor coil air.
 Low Temperature Heating Ratings — 17°F db/15°F wb outdoor air temperature and 70°F db entering indoor coil air.
*Heating Seasonal Performance Factor.
•Kit must be ordered extra and field installed.

FEATURES

Equipment Warranty — The compressor has a limited warranty for a full five years. All other components have a limited warranty for one year. Refer to Lennox Equipment Limited Warranty included with the unit for details.

Weather Resistant Cabinet — Heavy gauge galvanized steel cabinet is subject to a five station metal wash process. This preparation process results in a perfect bonding surface for the finish coat of baked-on enamel. The outdoor enamel paint finish gives the cabinet long lasting protection from the weather. Drainage holes are furnished in the base and base channels for moisture removal. Heavy duty channels under the base raise the unit off the mounting surface away from the damaging moisture. Control box is conveniently located for easy access with all controls installed and pre-wired at the factory.

Powerful Outdoor Fan — Efficient direct drive fan moves large volumes of air uniformly through the entire outdoor coil resulting in high refrigerant cooling capacity. Vertical discharge of air minimizes operating sounds and eliminates hot air damage to lawn and shrubs. Fan motor is totally enclosed for maximum protection from weather, dust and corrosion. A rain shield on the motor provides additional protection from moisture. Fan service access is accomplished by removal of fan guard. Corrosion resistant PVC coated steel wire fan guard is furnished as standard.

Refrigerant Line Connections, Electrical Inlets and Service Valves — Vapor and liquid line connections are located outside of the cabinet and are made with sweat connections. Brass service valves prevent corrosion and provide access to refrigerant system. Drier with internal check valve and strainer are factory installed in the liquid line. One-shot vapor valve, liquid line service valve and gauge ports are accessible outside of the cabinet. A field installed thermometer well is furnished. Refrigerant line connections, valve and field wiring inlets are all conveniently located in one central area of the cabinet. See dimension drawing.

Defrost Control — A factory installed clock timer defrost control is furnished as standard equipment. It gives a defrost cycle for every 45 or 90 minutes (adjustable) of compressor "on" time at outdoor temperature below 35°F. A sensing element mounted on the outdoor coil determines when the defrost cycle is required. Defrost pressure switch on the liquid line terminates cycle.

Expansion Valve — Designed and sized specifically for use in heat pump system. Sensor is located on the suction line between reversing valve and compressor thus sensing suction temperature in any cycle. Factory installed.

Reversing Valve — 4-way interchange reversing valve effects a rapid change in direction of refrigerant flow resulting in quick changeover from cooling to heating and vice versa. Valve operates on pressure differential between outdoor unit and indoor unit of the system. Factory installed.

Timed-Off Control — Furnished and factory installed. Provides low voltage protection and prevents compressor short-cycling. Automatic reset control provides a time delay between compressor shutoff and start-up.

Thermostat (Optional) — Thermostat is not furnished with the unit and must be ordered extra. See Accessories Section, Page 13 and Lennox Price Book.

Refrigerant Line Kits (Optional) — Lines are available in several lengths and must be ordered extra. See Refrigerant Line Kit table for selection. The refrigerant lines (vapor and liquid) are shipped refrigeration clean. Lines are cleaned, dried, pressurized and sealed at the factory. Vapor line is fully insulated. Lines are furnished with a flare fitting (Indoor unit connection) on one end and less any fitting (stubbed) on the opposite end for connection to the outdoor unit. Refrigerant line length should not exceed 50 ft. in any installation. If longer length lines are required, contact your Lennox Division Service Manager.

Mounting Base (Optional) — Rugged mounting base provides permanent foundation for outdoor units. High density polyethylene structural material is lightweight, sturdy, sound absorbing and will withstand the rigors of the sun, heat, cold, moisture, oil and refrigerant. Will not mildew or rot. Can be shipped singly or in packages of 6 to a carton. HP20-211, HP20-261, HP20-311 and HP20-411 models use the MB1-22 base (99C78) 22-1/4" x 22-1/4" x 3" shipping weight 10 lbs. HP20-461 model uses the MB1-32 base (83C83) 32" x 34" x 3" shipping weight 15 lbs.

Outdoor Thermostat Kit (Optional) — An outdoor thermostat can be used to lock out some of the electric heating elements on indoor units when two stage control is applicable. Outdoor thermostat maintains the heating load on the low power input as long as possible before allowing the full power load to come on the line. Thermostat kit (LB-29740BA) and mounting box (M-1595) must be ordered extra.

Check and Expansion Valve Kits (Optional) — Must be ordered extra and field installed on indoor units. See ARI Ratings table.

SPECIFICATIONS

	Model No.	HP20-211	HP20-261	HP20-311	HP20-411	HP20-461
Outdoor Coil	Net face area (sq. ft.)	9.24	7.39	7.39	9.24	11.39
	Tube diameter (in.)	3/8	3/8	3/8	3/8	3/8
	No. of rows	1	2	2	2	2
	Fins per inch	20	20	20	20	18
Outdoor Fan	Diameter (in.)	18	18	18	18	22
	No. of blades	4	4	4	4	4
	Motor hp	1/6	1/6	1/6	1/6	1/3
	Cfm	1900	2400	2400	2600	3800
	Rpm	1170	1040	1040	1060	1060
	Watts	150	250	250	260	400
	Refrigerant-22 (charge furnished)	4 lbs. 5 oz.	6 lbs. 1 oz.	6 lbs. 1 oz.	6 lbs. 15 oz.	8 lbs. 5oz.
	Liquid line connection (sweat)	5/16	5/16	3/8	3/8	3/8
	Vapor line connection (sweat)	5/8	5/8	3/4	3/4	7/8
	Shipping weight (lbs.)	138	148	148	170	217
	Number of packages in shipment	1	1	1	1	1

Blower Coil Data

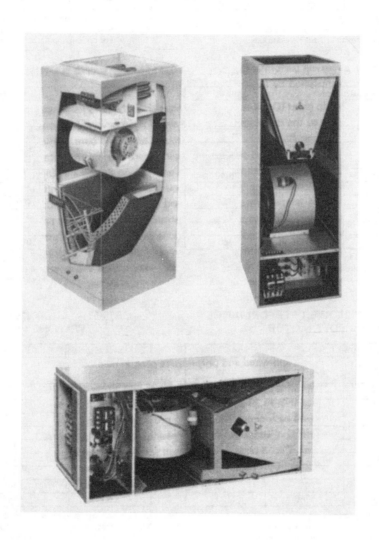

BLOWER DATA

CB19-21 AND CBH19-21 BLOWER PERFORMANCE

External Static Pressure (in. wg.)	Air Volume (cfm) @ Various Speeds		
	High	Medium	Low
0	855	630	530
.05	830	625	525
.10	800	620	520
.15	765	610	515
.20	730	595	505
.25	685	570	490
.30	640	535	465
.40	525	425	365

NOTE — All cfm is measured external to the unit.
NOTE — Electric heaters have no appreciable air resistance.
For optional up-flo air filter resistance see separate table.

CB19-26 AND CBH19-26 BLOWER PERFORMANCE

External Static Pressure (in. wg.)	Air Volume (cfm) @ Various Speeds		
	High	Medium	Low
0	1150	1020	870
.05	1105	985	860
.10	1065	955	850
.15	1020	920	825
.20	960	875	795
.25	905	830	755
.30	845	780	710
.40	680	625	550

NOTE — All cfm is measured external to the unit.
NOTE — Electric heaters have no appreciable air resistance.
For optional up-flo air filter resistance see separate table.

CB19-31 AND CBH19-31 BLOWER PERFORMANCE

External Static Pressure (in. wg.)	Air Volume (cfm) @ Various Speeds		
	High	Medium	Low
0	1400	1270	1050
.05	1370	1250	1050
.10	1335	1220	1050
.15	1290	1190	1040
.20	1240	1150	1025
.25	1190	1110	1000
.30	1130	1060	970
.40	1000	945	885
.50	855	815	765

NOTE — All cfm is measured external to the unit.
NOTE — Electric heaters have no appreciable air resistance.
For optional up-flo air filter resistance see separate table.

CB19-41 AND CBH19-41 BLOWER PERFORMANCE

External Static Pressure (in. wg.)	Air Volume (cfm) @ Various Speeds		
	High	Medium	Low
0	1630	1380	1130
.05	1590	1370	1150
.10	1550	1350	1160
.15	1500	1330	1160
.20	1450	1310	1160
.25	1400	1270	1150
.30	1340	1230	1130
.40	1200	1130	1050
.50	1010	960	890

NOTE — All cfm is measured external to the unit.
NOTE — Electric heaters have no appreciable air resistance.
For optional up-flo air filter resistance see separate table.

CB19-51 AND CBH19-51 BLOWER PERFORMANCE WITH 208/230 VOLT MOTOR

External Static Pressure (in. wg.)	Air Volume (cfm) @ Various Speeds		
	High	Medium	Low
0	1950	1640	1380
.05	1910	1620	1370
.10	1870	1600	1350
.15	1830	1580	1330
.20	1780	1550	1310
.25	1730	1520	1290
.30	1680	1490	1260
.40	1570	1400	1200
.50	1410	1280	1100

NOTE — All cfm is measured external to the unit.
NOTE — Electric heaters have no appreciable air resistance.
For optional up-flo air filter resistance see separate table.

CB19-51 AND CBH19-51 BLOWER PERFORMANCE WITH 460 VOLT (1 phase) MOTOR

External Static Pressure (in. wg.)	Air Volume (cfm) @ Various Speeds	
	High	Low
0	2020	1630
.05	1950	1610
.10	1920	1600
.15	1870	1570
.20	1820	1540
.25	1770	1500
.30	1710	1460
.40	1590	1350
.50	1430	1250

NOTE — All cfm is measured external to the unit.
NOTE — Electric heaters have no appreciable air resistance.
For optional up-flo air filter resistance see separate table.

CB19-65 AND CBH19-65 BLOWER PERFORMANCE WITH 208/230 VOLT MOTOR

External Static Pressure (in. wg.)	Air Volume (cfm) @ Various Speeds		
	High	Medium	Low
0	2415	2205	1830
.05	2360	2165	1815
.10	2305	2125	1800
.15	2245	2085	1780
.20	2185	2040	1760
.25	2130	2000	1735
.30	2070	1950	1705
.40	1940	1845	1630
.50	1810	1725	1540
.60	1665	1585	1405

NOTE — All cfm is measured external to the unit.
NOTE — Electric heaters have no appreciable air resistance.
For optional up-flo air filter resistance see separate table.

CB19-65 AND CBH19-65 BLOWER PERFORMANCE WITH 460 VOLT (1 phase) MOTOR

External Static Pressure (in. wg.)	Air Volume (cfm) @ Various Speeds	
	High	Low
0	2380	2250
.05	2340	2180
.10	2290	2140
.15	2250	2110
.20	2190	2065
.25	2130	2015
.30	2075	1970
.40	1945	1860
.50	1820	1760

NOTE — All cfm is measured external to the unit.
NOTE — Electric heaters have no appreciable air resistance.
For optional up-flo air filter resistance see separate table.

Blower-Coil Unit Model No.	Electric Heat Unit Model No. & Shipping Weight	No. of Steps & Phase	Volts Input	kw Input	Btuh Input	*Minimum Circuit Ampacity	
						Circuit 1	Circuit 2
CB19-21 CBH19-21	ECB19-2.5 (4 lbs.)	1 step 1 phase	208	1.9	6,400	12.3	----
			220	2.1	7,200	12.8	----
			230	2.3	7,800	13.4	----
			240	2.5	8,500	13.9	----
	ECB19-5 (4 lbs.)	1 step 1 phase	208	3.8	12,800	23.5	----
			220	4.2	14,300	24.8	----
			230	4.6	15,700	25.9	----
			240	5.0	17,100	26.9	----
	ECB19-6 (5 lbs.)	2 steps 1 phase	208	4.5	15,400	28.0	----
			220	5.0	17,100	29.3	----
			230	5.5	18,800	30.9	----
			240	6.0	20,500	32.2	----
	ECB19-7 (5 lbs.)	2 steps 1 phase	208	5.3	17,900	32.5	----
			220	5.9	20,100	34.4	----
			230	6.4	21,900	35.8	----
			240	7.0	23,900	37.4	----
	ECB19-8 (5 lbs.)	2 steps 1 phase	208	6.0	20,500	37.0	----
			220	6.7	22,900	39.0	----
			230	7.3	25,100	40.8	----
			240	8.0	27,300	42.6	----
	ECB19-10 (5 lbs.)	2 steps 1 phase	208	7.5	25,600	45.9	----
			220	8.4	28,700	48.6	----
			230	9.2	31,400	50.8	----
			240	10.0	34,100	53.0	----
CB19-26 CBH19-26	ECB19-2.5 (4 lbs.)	1 step 1 phase	208	1.9	6,400	13.2	----
			220	2.1	7,200	13.7	----
			230	2.3	7,800	14.3	----
			240	2.5	8,500	14.8	----
	ECB19-5 (4 lbs.)	1 step 1 phase	208	3.8	12,800	24.4	----
			220	4.2	14,300	25.7	----
			230	4.6	15,700	26.8	----
			240	5.0	17,100	27.8	----
	ECB19-6 (5 lbs.)	2 steps 1 phase	208	4.5	15,400	28.9	----
			220	5.0	17,200	30.2	----
			230	5.5	18,800	31.8	----
			240	6.0	20,500	33.7	----
	ECB19-7 (5 lbs.)	2 steps 1 phase	208	5.3	17,900	33.4	----
			220	5.9	20,100	35.3	----
			230	6.4	21,900	36.7	----
			240	7.0	23,900	38.3	----
	ECB19-8 (5 lbs.)	2 steps 1 phase	208	6.0	20,500	37.9	----
			220	6.7	22,900	39.9	----
			230	7.3	25,100	41.7	----
			240	8.0	27,300	43.4	----
	ECB19-10 (5 lbs.)	2 steps 1 phase	208	7.5	25,600	46.8	----
			220	8.4	28,700	49.5	----
			230	9.2	31,400	51.7	----
			240	10.0	34,100	53.9	----
	ECB19-12.5 (10 lbs.)	3 steps 1 phase	208	9.4	32,000	39.4	18.9
			220	10.5	35,800	41.2	19.9
			230	11.5	39,200	43.4	20.8
			240	12.5	42,600	45.2	21.8
	ECB19-15 (10 lbs.)	3 steps 1 phase	208	11.3	38,400	46.9	22.7
			220	12.6	43,000	49.1	23.9
			230	13.5	47,000	51.7	25.0
			240	15.0	51,200	53.9	26.0

*Refer to National Electrical Code manual to determine wire, fuse and disconnect size requirements. Use wires suitable for at least 167°F.

FEATURES

Completely Tested — Blower coil units are tested with matching condensing units in the Lennox Research Laboratory environmental test room in accordance with ARI Standard 210-81. Optional electric heaters are rated in accordance with Department of Energy (DOE) test procedures and Federal Trade Commission (FTC) labeling regulations. Blower performance data is according to actual unit tests conducted in Lennox air test chamber. Blower-coil units and components within are bonded for grounding to meet safety standards for servicing required by U.L. and NEC.

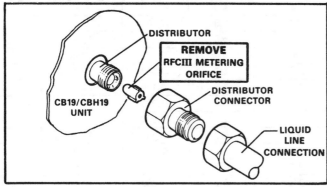

Expansion Valve Kits (Optional) — Expansion valve field installs on the blower coil unit external to the cabinet. The blower coil unit must be field altered by removing the RFCIII metering orifice, see sketch above. Expansion valve kits are optional and must be ordered extra. See kit table for requirement.

Durable Cabinet — Constructed of heavy gauge galvanized steel and completely insulated with thick fiberglass insulation. The cabinets are subject to a five station metal wash process before painting. This preparation process results in a perfect bonding surface for the attractive finish coat of baked-on enamel. Removable panels provide complete service access. Electrical inlets are provided in both sides of cabinet. Return air entry is possible in either side or bottom of cabinet on up-flo units.

Copper Tube/Enhanced Fin Indoor Coil — Lennox designed twin coils, assembled in a 'V' configuration, provides extra large surface and contact area, excellent heat transfer and low air resistance for maximum efficiency. Precise circuiting gives uniform refrigerant distribution. Lennox fabricated coil is constructed of precisely spaced ripple-edged aluminum fins fitted to durable seamless copper tubes. Fins are strengthened to resist bending and are equipped with collars that grip tubing for maximum contact area. Lanced fins provide maximum exposure of fin surface to air stream. Flared shoulder tubing joints and silver soldering provide tight, leakproof joints. Long life copper tubing is easy to field service. Coil is thoroughly factory tested under high pressure to insure leakproof construction.

Drain Pan — Deep, corrosion resistant drain pan has dual pipe drains extended outside of cabinet for ease of connection. See dimension drawings.

Refrigerant Line Connections — Suction and liquid lines are equipped with flare fittings and extend outside of the cabinet for ease of connection. CB19/CBH19-65 suction line requires sweat connection. See dimension drawings for location.

Transformer and Blower Cooling Relay (Furnished) — A 2 volt transformer and blower cooling relay are furnished as st ard equipment and are factory installed in the unit control b A terminal strip is also furnished as standard.

Powerful Blower — Equipped with a Lennox designed and buil direct drive blower. Each blower is statically and dynamicall balanced as an assembly before it is installed in the unit. Multispee motor is resiliently mounted. A choice of blower speeds is available See blower performance charts. Change of blower speeds is easil accomplished by a simple change in wiring.

Electric Heat (Optional) — Additive electric heaters field insta internal to the unit cabinet and are available in several kw sizes see Electric Heat table. The helix wound nichrome bare heatin elements are exposed directly in the air stream resulting in instan heat transfer, low element temperatures and long service life. Eac heating element is equipped with accurately located limit contr with fixed temperature off setting and automatic reset. In addi tion, elements have supplemental thermal cutoff safety fuses pro viding positive protection in case of excessive temperatures. Cutof fuses are mounted external to the element face plate for quick an easy replacement. Thermal sequencer relay brings the heatin elements on and off line, in sequence and equal increments, wit a time delay between each element. Sequencer also initiates an terminates blower operation. Heating control relay(s), is furnishe as standard. Control box and access cover are constructed of heav gauge galvanized steel. Heaters are factory assembled with con trols installed and wired and only require plug-in field connec

Circuit Breakers — ECB19-12.5,15,20,25 and 30 kw (208/240v 1ph) and ECB19-15,20 and 25 kw (208/240v-3ph) electric heater are equipped with circuit breakers to provide overload and sho circuit protection. Breakers are factory wired and mounted on elec tric heat unit. Circuit breakers are current sensitive and temperatur actuated to shut off heater if current draw is excessive. Must b reset manually. Circuit breakers qualify as the disconnect mean at unit in many areas and eliminate the need for a field provide disconnect. Consult local electrical code in your area.

Up-Flo Side Return Air Filter Adapter (Optional) — Field in stalls on either side return air opening of up-flo cabinet. Constructe of heavy gauge galvanized steel with a baked-on paint finish. Equip ped with flanges for ease of duct connection. CB19-51 and CB19-6 adapter is insulated with thick fiberglass insulation. Access pane allows easy removal and replacement of filter(s). One inch thic frame type filter is furnished as standard. Media is washable o vacuum cleanable oil coated polyurethane. CB19-21 thru 41 model use one filter. CB19-51 and CB19-65 models have two. See spec fication table for sizes. CB19-51 and CB19-65 adapter is shippe knocked down and must be field assembled.

Down-Flo Additive Base (Optional) — An optional additiv base is required for models with electric heat installed in the down flo position on combustible floors. Base is not furnished and mus be ordered extra for field installation. See Specifications tabl dimension drawing.

Water-to-Air Heat Pump Equipment

EWT	GPM	WPD	COOLING						HEATING				
			EA	TC	SC	KW	HR	EER	EA	HC	KW	HE	COP
30	4.5	3.5	75/63 80/67 85/71	40.3 43.8 47.3	27.8 29.0 30.2	2.46 2.51 2.56	48.7 52.3 56.0	16.3 17.4 18.4	60 70 80	26.8 25.9 24.9	2.45 2.58 2.71	18.4 17.1 15.7	3.20 2.94 2.70
	7	8.4	75/63 80/67 85/71	41.5 45.1 48.7	28.4 29.6 30.8	2.39 2.44 2.49	49.7 53.4 57.2	17.3 18.5 19.6	60 70 80	28.0 27.0 26.0	2.52 2.66 2.79	19.4 18.0 16.5	3.25 2.98 2.73
	9	13.5	75/63 80/67 85/71	42.8 46.5 50.2	29.0 30.2 31.4	2.32 2.37 2.42	50.7 54.6 58.5	18.4 19.6 20.8	60 70 80	29.3 28.3 27.2	2.60 2.74 2.88	20.4 18.9 17.4	3.30 3.02 2.77
50	4.5	3.0	75/63 80/67 85/71	38.7 42.1 45.4	27.6 28.7 29.9	2.65 2.71 2.76	47.7 51.3 54.8	14.6 15.5 16.5	60 70 80	36.2 35.0 33.7	2.87 3.02 3.17	26.4 24.6 22.9	3.69 3.39 3.12
	7	7.1	75/63 80/67 85/71	39.9 43.4 46.8	28.1 29.3 30.5	2.57 2.63 2.68	48.7 52.3 56.0	15.5 16.5 17.5	60 70 80	37.9 36.6 35.3	2.96 3.12 3.27	27.8 25.9 24.1	3.75 3.44 3.16
	9	11.4	75/63 80/67 85/71	41.1 44.7 48.3	28.7 29.9 31.1	2.50 2.55 2.60	49.7 53.4 57.2	16.5 17.5 18.6	60 70 80	39.5 38.2 36.8	3.02 3.18 3.34	29.2 27.3 25.4	3.83 3.52 3.23
70	4.5	2.8	75/63 80/67 85/71	34.3 37.3 40.2	25.5 26.6 27.7	3.04 3.10 3.16	44.6 47.8 51.0	11.3 12.0 12.7	60 70 80	46.3 44.9 42.8	3.30 3.47 3.64	35.0 33.0 30.4	4.11 3.79 3.44
	7	6.6	75/63 80/67 85/71	35.3 38.4 41.5	26.1 27.1 28.2	2.95 3.01 3.07	45.4 48.7 52.0	12.0 12.8 13.5	60 70 80	48.5 47.0 45.5	3.40 3.58 3.76	36.9 34.8 32.7	4.18 3.85 3.55
	9	10.6	75/63 80/67 85/71	36.4 39.6 42.8	26.6 27.7 28.8	2.86 2.92 2.98	46.2 49.6 52.9	12.7 13.6 14.4	60 70 80	50.6 49.1 47.5	3.47 3.65 3.83	38.8 36.6 34.4	4.28 3.94 3.63
90	4.5	2.6	75/63 80/67 85/71	29.9 32.5 35.1	22.7 23.6 24.6	3.39 3.46 3.53	41.4 44.3 47.1	8.8 9.4 9.9	60 70 80	53.2 51.9 50.7	3.66 3.85 4.04	40.7 38.8 36.9	4.26 3.95 3.67
	7	6.2	75/63 80/67 85/71	30.8 33.5 36.1	23.1 24.1 25.1	3.29 3.36 3.42	42.0 44.9 47.8	9.4 10.0 10.6	60 70 80	55.3 54.0 52.6	3.77 3.97 4.17	42.4 40.4 38.4	4.30 3.98 3.70
	9	10.0	75/63 80/67 85/71	31.7 34.5 37.3	23.6 24.6 25.6	3.19 3.26 3.33	42.6 45.6 48.6	9.9 10.6 11.2	60 70 80	57.3 55.9 54.5	3.85 4.05 4.25	44.2 42.1 40.0	4.37 4.05 3.76
110	4.5	2.5	75/63 80/67 85/71	25.6 27.9 30.1	19.9 20.7 21.6	3.74 3.82 3.90	38.4 40.9 43.4	6.8 7.3 7.7					
	7	5.9	75/63 80/67 85/71	26.4 28.7 31.0	20.3 21.2 22.0	3.63 3.71 3.78	38.8 41.4 43.9	7.3 7.7 8.2					
	9	9.5	75/63 80/67 85/71	27.2 29.6 32.0	20.7 21.6 22.5	3.53 3.60 3.67	39.3 41.9 44.5	7.7 8.2 8.7					

EWT	GPM	WPD	COOLING							HEATING					
			EA	TC	SC	KW	HR	DC	EER	EA	HC	KW	HE	DC	COP
30	4.5	3.5	75/63 80/67 85/71	40.7 44.2 47.7	28.1 29.3 30.5	2.43 2.48 2.53	46.6 50.2 53.7	2.4 2.5 2.7	16.7 17.8 18.9	60 70 80	25.6 23.1 20.6	2.45 2.53 2.61	18.4 17.2 16.0	1.2 2.8 4.3	3.20 2.99 2.80
	7	8.4	75/63 80/67 85/71	41.9 45.5 49.1	28.7 29.9 31.1	2.36 2.41 2.46	47.9 51.5 55.2	2.0 2.2 2.3	17.8 18.9 20.0	60 70 80	26.6 24.1 21.5	2.52 2.60 2.69	19.4 18.2 16.9	1.4 3.0 4.5	3.26 3.04 2.84
	9	13.5	75/63 80/67 85/71	43.1 46.9 50.6	29.2 30.5 31.7	2.29 2.33 2.38	49.2 52.9 56.7	1.8 1.9 2.1	18.9 20.1 21.2	60 70 80	27.7 25.1 22.5	2.59 2.68 2.77	20.5 19.1 17.8	1.7 3.2 4.7	3.32 3.09 2.88
50	4.5	3.0	75/63 80/67 85/71	39.3 42.7 46.1	28.0 29.2 30.3	2.62 2.68 2.73	44.2 47.6 51.0	4.1 4.3 4.4	15.0 16.0 16.9	60 70 80	33.3 30.5 27.8	2.82 2.92 3.03	26.6 25.0 23.4	2.9 4.4 5.9	3.76 3.50 3.26
	7	7.1	75/63 80/67 85/71	40.5 44.0 47.5	28.6 29.7 30.9	2.54 2.60 2.65	45.4 48.9 52.4	3.8 4.0 4.1	15.9 16.9 17.9	60 70 80	34.6 31.9 29.1	2.90 3.01 3.12	28.0 26.3 24.6	3.2 4.7 6.2	3.83 3.56 3.32
	9	11.4	75/63 80/67 85/71	41.7 45.3 48.9	29.1 30.3 31.5	2.47 2.52 2.57	46.6 50.2 53.8	3.5 3.7 3.9	16.9 18.0 19.0	60 70 80	36.0 33.1 30.3	2.95 3.06 3.18	29.5 27.7 25.9	3.5 5.0 6.5	3.93 3.65 3.39
70	4.5	2.8	75/63 80/67 85/71	35.1 38.1 41.1	26.2 27.2 28.3	3.01 3.07 3.14	39.6 42.7 45.8	5.8 5.9 6.1	11.7 12.4 13.1	60 70 80	41.5 38.6 35.2	3.19 3.31 3.44	35.4 33.6 31.0	4.8 6.2 7.6	4.25 3.97 3.64
	7	6.6	75/63 80/67 85/71	36.2 39.2 42.3	26.7 27.7 28.8	2.92 2.98 3.04	40.6 43.8 47.0	5.5 5.6 5.7	12.4 13.2 13.9	60 70 80	43.3 40.3 37.4	3.28 3.41 3.54	37.3 35.3 33.2	5.2 6.6 8.1	4.33 4.04 3.76
	9	10.6	75/63 80/67 85/71	37.2 40.4 43.6	27.2 28.3 29.4	2.84 2.90 2.95	41.7 44.9 48.2	5.2 5.4 5.5	13.1 14.0 14.8	60 70 80	45.1 42.0 39.0	3.33 3.47 3.61	39.3 37.2 35.1	5.6 7.0 8.4	4.45 4.14 3.85
90	4.5	2.6	75/63 80/67 85/71	30.9 33.5 36.1	23.5 24.4 25.3	3.37 3.44 3.51	34.9 37.7 40.4	7.5 7.6 7.7	9.2 9.7 10.3	60 70 80	47.2 44.4 41.6	3.51 3.65 3.80	41.2 39.5 37.7	6.0 7.5 9.0	4.44 4.16 3.90
	7	6.2	75/63 80/67 85/71	31.8 34.5 37.2	23.9 24.9 25.8	3.27 3.34 3.41	35.8 38.6 41.4	7.2 7.3 7.4	9.7 10.3 10.9	60 70 80	48.9 46.0 43.2	3.61 3.76 3.91	43.0 41.1 39.2	6.4 7.9 9.4	4.49 4.20 3.94
	9	10.0	75/63 80/67 85/71	32.7 35.5 38.3	24.3 25.3 26.3	3.17 3.24 3.31	36.7 39.6 42.4	6.9 7.0 7.1	10.3 11.0 11.6	60 70 80	50.5 47.6 44.8	3.67 3.83 3.99	44.8 42.8 40.9	6.8 8.3 9.7	4.57 4.28 4.00
110	4.5	2.5	75/63 80/67 85/71	26.9 29.1 31.4	20.9 21.7 22.5	3.73 3.81 3.88	30.4 32.8 35.2	9.2 9.3 9.4	7.2 7.7 8.1						
	7	5.9	75/63 80/67 85/71	27.6 29.9 32.3	21.3 22.1 22.9	3.62 3.69 3.77	31.2 33.6 36.1	8.8 8.9 9.0	7.6 8.1 8.6						
	9	9.5	75/63 80/67 85/71	28.4 30.8 33.2	21.6 22.5 23.3	3.51 3.59 3.66	31.9 34.4 36.9	8.6 8.7 8.8	8.1 8.6 9.1						

Blower Performance Data

Includes allowances for wet coil and for filter

UNIT SIZE	BLOWER SPEED	EXTERNAL STATIC PRESSURE									
		0.10	0.15	0.20	0.25	0.30	0.35	0.40	0.50	0.60	0.70
WX009	LOW	320	310	300	290	275	—	—	—	—	—
WX009	HIGH	380	370	360	350	335	—	—	—	—	—
WX012	LOW	350	335	320	305	290	—	—	—	—	—
WX012	HIGH	400	385	365	350	335	—	—	—	—	—
WX019	LOW	560	550	540	530	520	—	—	—	—	—
WX019	HIGH	665	650	635	620	605	—	—	—	—	—
WX025	LOW	845	825	805	780	750	—	—	—	—	—
WX025	HIGH	895	875	850	825	795	765	—	—	—	—
WX031	LOW	1000	985	970	950	920	885	—	—	—	—
WX031	HIGH	1040	1030	1010	995	970	940	910	—	—	—
WX033	LOW	1010	980	960	930	890	850	—	—	—	—
WX033	HIGH	1080	1050	1010	980	940	900	870	—	—	—
WX036	LOW	1125	1110	1090	1070	1050	1030	—	—	—	—
WX036	HIGH	1200	1180	1160	1145	1115	1090	1060	—	—	—
WX041	LOW	1215	1210	1205	1200	1190	1180	1170	—	—	—
WX041	HIGH	1395	1385	1375	1365	1350	1335	1310	1260	—	—
WX049	LOW	1320	1315	1310	1300	1290	1275	1260	1220	—	—
WX049	HIGH	1650	1625	1600	1575	1545	1510	1475	1425	1340	—
WX059	LOW	1765	1750	1735	1720	1700	1680	1655	1600	1550	—
WX059	HIGH	1960	1930	1905	1880	1855	1825	1790	1730	1670	1600

All units factory wired for high speed.

Physical Data

UNIT SIZE	WX009	WX012	WX019	WX025	WX031	WX033	WX036	WX041	WX049	WX059
FAN WHEEL DIA x WIDTH (IN.)	6 x 6	6 x 8	9 x 6	9 x 7	9 x 7	9 x 7	10 x 6	10 x 6	10 x 6	11 x 8
FAN MOTOR HORSEPOWER	1/15	1/10	1/8	1/6	1/3	1/3	1/3	1/3	1/2	3/4
AIR COIL FACE AREA (SQ. FT.)	1.1	1.1	2.3	2.3	2.3	2.3	2.3	4.9	4.9	4.9
AIR COIL NO. OF ROWS	2	3	3	4	3	5	3	3	4	4
REFRIGERANT CHARGE R22 (OZ.)	18.0	18.0	28.0	39.0	46.0	50.0	55.0	85.0	97.0	97.0
FILTER-1" THICK HORIZONTAL	10 x 16		20 x 20					19.6 x 35.4		
FILTER-1" THICK VERTICAL	10 x 16		20 x 20					30.8 x 22.0		
UNIT WEIGHT (LBS.) HORIZONTAL	100	104	185	200	230	240	250	290	320	335
UNIT WEIGHT (LBS.) VERTICAL	102	106	190	205	235	245	255	295	325	340

Electrical Data

UNIT SIZE	60 HZ POWER		COMPRESSOR		FAN MOTOR FLA	TOTAL UNIT FLA	MINIMUM VOLTS	MINIMUM CIRCUIT AMPACITY	MAXIMUM FUSE SIZE
	VOLTS	PHASE	RLA	LRA					
WX009	208-230	1	4.1	20.0	0.52	4.6	197	5.6	10
	265	1	3.1	16.0	0.50	3.6	239	4.4	10
WX012	208-230	1	6.0	31.0	0.65	6.7	197	8.2	10
	265	1	4.5	27.0	0.56	5.1	239	6.2	10
WX019	208-230	1	8.2	49.0	1.0	9.2	197	11.3	15
	265	1	7.3	44.0	0.9	8.2	239	10.0	15
WX025	208-230	1	12.0	64.0	1.0	13.0	197	16.0	25
	265	1	11.1	61.0	0.9	12.0	239	14.8	25
	208-230	3	7.7	50.0	1.0	8.7	197	10.6	15
WX031	208-230	1	13.3	80.0	2.1	15.4	197	18.7	30
	265	1	11.9	72.0	1.8	13.7	239	16.7	25
	208-230	3	8.4	80.0	2.1	10.5	197	12.6	20
WX033	208-230	1	13.7	87.0	2.1	15.8	197	19.2	30
	208-230	3	9.3	71.0	2.1	11.4	197	13.7	20
	460	3	4.7	36.0	1.1	5.8	414	7.0	10
WX036	208-230	1	13.4	78.8	2.2	15.6	197	19.0	30
	208-230	3	8.9	59.5	2.2	11.1	197	13.3	20
	460	3	4.4	30.7	1.2	5.6	414	6.6	10
WX041	208-230	1	18.2	98.0	2.2	20.4	197	25.0	40
	208-230	3	11.8	73.0	2.2	14.0	197	17.0	25
	460	3	5.9	38.0	1.2	7.1	414	8.6	10
WX049	208-230	1	21.2	110.0	3.3	24.5	197	29.8	50
	208-230	3	13.7	92.0	3.3	17.0	197	20.4	30
	460	3	6.4	46.0	2.0	8.4	414	10.0	15
WX059	208-230	1	24.6	148.0	4.3	28.9	197	35.1	50
	208-230	3	13.6	137.0	4.3	17.9	197	21.3	30
	460	3	6.2	70.0	2.3	8.5	414	10.1	15

GENERAL — The liquid source heat pumps shall be either suspended type with horizontal air inlet and discharge, or floor mounted type with horizontal air inlet and vertical air discharge reverse cycle heating/cooling units. Units shall be A.R.I. performance certified, and listed by a nationally recognized safety testing laboratory, or agency, such as Electrical Testing Laboratory (E.T.L.), or Canadian Standards Association (C.S.A.). Each unit shall be computer run-tested at the factory. Each unit shall be shipped in a corrugated box.

The units shall be warranted by the manufacturer against defects in materials and workmanship for a period of one year. An optional four year extended warranty on the motor compressor unit and major refrigerant circuit components shall be available.

The liquid source heat pump units shall be designed to operate with entering liquid temperature between 20°F and 110°F as manufactured by WaterFurnace International of Fort Wayne, Indiana.

CASING AND CABINET — The cabinet shall be fabricated from heavy-gauge steel and finished with corrosion resistant textured epoxy coating. The interior shall be insulated with 1/2" thick, multi-density, coated glass fiber with edges sealed or tucked under flanges to prevent the introduction of glass fibers into the discharge air. One blower and two compressor compartment access panels shall be removable with supply and return ductwork in place. A duct collar shall be provided on the supply air opening. A 1" thick throwaway glass fiber filter shall be provided with each unit. Insulated galvanized steel condensate drain pans shall be provided and extend beyond the coil to catch blow off. The units shall have an insulated divider panel between the air handling section and the compressor-control section to minimize the transmission of compressor noise, and to permit operational service testing through either compressor access panel without having air by-pass the refrigerant to air coil.

Vertical units can be supplied in left-hand return air or front (can be used right) return air configurations (as viewed from the water and electrical connection side).

All units shall have 7/8" knockouts for entrance of low and line voltage wiring.

REFRIGERANT CIRCUIT — All units shall contain a sealed refrigerant circuit including a hermetic motor-compressor, bi-directional capillary tube/thermal expansion valve assembly, finned-tube air-to-refrigerant heat exchanger, reversing valve, coaxial tube water-to-refrigerant heat exchanger, factory installed high and low pressure safety switches and service ports for connection of high and low pressure gauges.

Compressors shall be designed for heat pump duty with internal isolation and mounted on vibration isolators. Compressor motors shall have overload protection and will be three-phase or single-phase PSC type. The finned-tube coil shall be constructed of rippled and corrugated aluminum fins bonded to seamless copper tubes in a staggered pattern.

The water-to-refrigerant heat exchanger shall be a coaxial type constructed of a convoluted copper (optional cupronickel) inner tube and a steel outer tube capable of withstanding 450 PSIG working pressure on the refrigerant side. The parallel capillary tube/thermal expansion valve assembly shall provide proper superheat over the 20°F-110°F liquid temperature range with minimal "hunting". The assembly shall operate bi-directionally without the use of check valves.

The water-to-refrigerant heat exchanger, optional desuperheater coil, and refrigerant suction lines shall be insulated to prevent condensation at low liquid temperatures.

FAN AND MOTOR ASSEMBLY — The fan shall be a direct drive centrifugal type. The fan wheel shall be dynamically balanced. On unit sizes 009 and 012, the fan housing shall have a removable end ring for ease of fan wheel removal. On unit sizes 019 through 059, the fan housing shall be disconnectable from the unit without removing the supply air ductwork for servicing of the fan motor. The fan motor shall be a multi-speed, PSC type. The fan motor shall be isolated from the housing by rubber grommets. The motor shall be permanently lubricated and have thermal overload protection. On horizontal units, the fan discharge shall be supplied as end discharge or straight discharge with the capability of field conversion from one to the other.

ELECTRICAL — Controls and safety devices will be factory wired and mounted within the unit. Controls shall include fan relay, compressor contactor, 24V transformer, reversing valve coil and reset relay. A terminal block with screw terminals will be provided for field control wiring. On all vertical units and on horizontal unit sizes 019 through 033, the control box can be rotated for access through the adjacent access panel. To prevent short cycling, the reset relay shall provide a lockout circuit when the safety controls are activated which requires resetting at the thermostat or main circuit breaker. A lockout indicating signal shall be provided on the low voltage terminal block.

PIPING — Condensate, supply, and return water connections (and optional desuperheater connections) shall be copper threaded fittings mechanically fastened to the unit cabinet, eliminating the need for backup wrenches when making field piping connections. All water piping shall be insulated to prevent condensation at low liquid temperatures.

HANGER KIT (Field Installed — Horizontal Units Only) — The hanger kit shall consist of four (4) galvanized steel brackets, bolts, lockwashers, and isolators and shall be designed to fasten to the unit bottom panel for suspension from 3/8" threaded rods.

ACCESSORIES AND OPTIONS

THERMOSTAT (Field Installed) — One of the following room thermostats shall be provided: (1) A standard manual changeover room thermostat with a subbase for manual selection of "HEAT-OFF-COOL" system operation and "ON-AUTO" blower operation; or (2) A standard automatic changeover room thermostat with a subbase for manual selection of "AUTO-OFF" system operation and "ON-AUTO" blower operation. Thermostat shall have separate levers for heating and cooling set points. (3) A deluxe manual changeover room thermostat with a built-in 8°F night heating temperature setback

initiated by the setback relay (timeclock). Subbase switching is similar to the standard MCO thermostat with the addition of a "normal-override" mode switch. (4) A deluxe automatic changeover room thermostat with built-in 8°F night heating temperature setback initiated by the setback relay (timeclock). Set point levers and subbase switching are similar to the standard ACO thermostat with the addition of a "normal-override" mode switch. (5) A manual changeover auxiliary heat thermostat with two stages of heating, "HEAT-OFF-COOL" system operation, "ON-AUTO" blower operation and "HP-NORM-EM" mode switches. "AUX HEAT and EM HEAT" indicating LED's are provided.

DESUPERHEATER — An optional heat reclaiming desuperheater coil of vented double wall copper construction suitable for potable water shall be provided.

RETURN AIR DUCT COLLAR (Field Installed) — A specially designed duct kit shall provide for a return air duct connection. The kit shall include 1" standard size filters.

ELECTROSTATIC FILTER — A permanent cleanable 93% efficient electrostatic filter may be provided in lieu of the standard throwaway type.

RANDOM START RELAY (Field Installed) — A time delay relay shall provide random delayed start-up from 1 to 60 seconds. The relay may be factory wired.

NIGHT SET BACK RELAY (Field Installed) — A relay shall shut down the unit from a separate 24V signal. The relay may be factory wired.

THERMOSTAT GUARD (Field Installed) — A replacement thermostat cover with keylock and external thermometer provides a tamperproof feature.

EARTH LOOP PUMP KIT (Field Installed) — A specially designed module shall provide all liquid flow, fill, and connection requirements for independent single unit systems.

DESUPERHEATER PUMP KIT (Field Installed) — A pump kit provides for connecting desuperheater coil in heat pump with water storage tank. In line fusing shall be provided.

AUXILIARY ELECTRIC HEATER — A duct mounted electric heater shall provide supplemental and/or emergency heating capability when used with the optional auxiliary heat thermostat.

FREE COOLING COIL MODULE (Field Installed) — A chilled water finned-tube coil shall provide "free" cooling when liquid temperatures allow. The coil module shall contain an adjustable temperature control and flow switching valves for fully automatic operation.

Filter Data

AIR RESISTANCE

Air Volume (cfm)	Total Resistance (in. wg)	
	EAC11-14	EAC11-20
800	.04	.03
1000	.05	.04
1200	.07	.05
1400	.09	.07
1600	----	.09
1800	----	.12
2000	----	.14

NOTE — Standard central system filter is removed and not included in table.

AIR CLEANING EFFICIENCY

On the average, an electronic air cleaner will remove fifteen (15) times as much dust, dirt, lint and mold spores from the air as an ordinary furnace filter. And, on smaller particles, the percentage removed vs. standard filters is significantly greater.

An electronic air cleaner will remove airborne particles as small as 0.01 microns in diameter. The chart below lists sizes of common airborne particles trapped and removed from recirculated air by electronic air cleaners.

Types Of Airborne Particles	Particle Size (*Microns)
Pollen	10.0 to 100.0
Tobacco Smoke	0.01 to 1.0
Cooking Smoke	0.02 to 1.0
Household Dust	0.01 to 300.0
Mold Spores	10.0 to 30.0
Atmospheric Dust	0.01 to 1.0
Insecticide Dust	0.40 to 10.0
Coal Dust (Soot)	1.0 to 100.0

*One micron = 1/25,400th of an inch.
 Particles 10 microns and larger are visible to naked eye.
 Particles 10 to 0.1 microns are visible with microscope.
 Particles below 0.1 microns are visible with electron microscope.

SPECIFICATIONS

Model No.	MF1-20
Air volume range (cfm)	600 — 2000
Filter size (in.) H x W x D	20 x 24-1/4 x 5
Shipping weight (lbs.)	16
No. of packages in shipment	1

AIR RESISTANCE

Air Volume (cfm)	Total Resistance (in. wg)
600	.04
800	.05
1000	.09
1200	.12
1400	.15
1600	.18
1800	.22
2000	.27

NOTE — Standard central system filter is removed and not included in table.

APPENDIX 3
Performance Models for Refrigeration Cycle Equipment

A3-1 Performance Models

The manufacturer's application data can be reduced to a set of linear equations. These equations are useful because they provide a generic model of cooling equipment performance. These models can be used to evaluate equipment performance (approximately) when comprehensive capacity tables are not available.

Performance models also are useful to software developers because equations simplify the programming work, reduce memory requirements and eliminate interpolation and extrapolation subroutines. The methodology for developing a set of equations that describes the performance of residential cooling equipment is presented in the following sections.

A3-2 Air-to-Air Cooling Model

The performance of air-to-air cooling equipment is sensitive to four parameters:

- Outdoor dry-bulb temperature (OAT)
- Entering dry-bulb temperature (EDB)
- Entering wet-bulb temperature (EWB)
- CFM flowing through the refrigerant coil (CFM)

These parameters can be combined into a set of linear equations that provide the same information contained in the manufacturer's performance data tables. These equations are presented below in their most general form. In this case, the first equation yields a value for the total capacity (TC), the second equation provides information about the sensible-to-total capacity ratio (SHR), and the third equation estimates the power requirement (Kw).

$$TC = K + M1 \times CFM + M2 \times EWB + M3 \times OAT$$

$$SHR = Q + N1 \times CFM + N2 \times EWB + N3 \times EDB + N4 \times OAT$$

$$Kw = R + O1 \times CFM + O2 \times EWB + O3 \times OAT$$

When a set of comprehensive performance data is available, the intercepts (K, Q, and R) and slopes (Mi, Ni and Oi) of the preceding equations can by evaluated by performing a multiple regression analysis on the entire set of data that is associated with a particular product or a group of products. Preliminary efforts along this line indicate that the Mi and Ni slopes are not too sensitive to the equipment capacity or the SEER rating. However the Oi slopes do depend on the SEER

rating. It also seems that the Q intercept is not sensitive to equipment capacity.

Table A3-1 provides a set of generic slopes and a Q-intercept for air-to-air cooling equipment. These values were generated by analyzing the performance data that was published by four different manufacturers (heat pumps and cooling-only equipment that was manufactured between 1980 and 1986).

Generic Slopes and Intercepts Air-to-Air Cooling		
M Slopes	N Slopes	O Slopes
M1 = 3.33	N1 = 0.0002	O1 = 0.20
M2 = 500	N2 = -0.0475	O2 = 27.5
M3 = -225	N3 = 0.0325	O3 = 24.5
	N4 = 0.0025	
Q Intercept = 0.82		

Table A3-1

A3-3 Example — Air-to-Air Extrapolation

A model that approximates the performance for a particular air-to-air equipment package can be generated from the ARI certification data if a set of generic slopes (Mi, Ni and Oi) and a generic Q intercept are available. For example, the values that are summarized by Table A3-1 will be applied to the following ARI rating data.

- TC = 37,400 BTUH
- CFM = 1,350
- SEER = 9.74
- OAT = 95 °F
- EWB = 67 °F
- EDB = 80 °F

Step 1 - Estimate the ARI power draw.

$$ARI \text{ Watts} = \frac{37400}{9.74} = 3840$$

Step 2 - Evaluate the K and R intercepts

$$37400 = K + (3.33 \times 1350) + (500 \times 67) - (225 \times 95)$$

$$K = 20780$$

$$3840 = R + (0.20 \times 1350) + (27.5 \times 67) + (24.5 \times 95)$$

$$R = 600$$

Step 3 - Write the Performance Equations

When the performance equations are assembled, they will have the following form:

$$TC = 20780 + (3.33 \times CFM) + (500 \times EWB) - (225 \times OAT)$$

$$SHR = 0.82 + (0.0002 \times CFM) - (0.0475 \times EWB) +$$

$$(0.0325 \times RAT) + (0.0025 \times OAT)$$

$$Watts = -600 + (0.20 \times CFM) + (27.5 \times EWB) + (24.5 \times OAT)$$

A3-4 Water-to-Air Cooling Model

The cooling performance of water-to-air equipment is sensitive to five parameters:

• Temperature of the water entering the condenser (EWT)
• Entering dry-bulb temperature (EDB)
• Entering wet-bulb temperature (EWB)
• CFM flowing through the refrigerant coil (CFM)
• GPM of water flowing through the condenser

One of these parameters (the GPM) can be ignored because the cooling performance is not too sensitive to the water-side flow rate. This leaves four parameters that can be combined into a set of linear equations that provide the same information that is contained in the manufacturers performance data tables. These equations are presented below in their most general form. In this case, the first equation yields a value for the total capacity (TC), the second equation provides information about the sensible-to-total capacity ratio (SHR) and the third equation estimates the power requirement (Kw).

$$TC = K + M1 \times CFM + M2 \times EWB + M3 \times EWT$$

$$SHR = Q + N1 \times CFM + N2 \times EWB + N3 \times EDB + N4 \times EWT$$

$$Kw = R + O1 \times CFM + O2 \times EWB + O3 \times EWT$$

When a set of comprehensive performance data is available, the intercepts (K, Q and R) and slopes (Mi, Ni and Oi) of the above equations can by evaluated by performing a multiple regression analysis on the entire set of data that is associated with a particular product or a group of products. Preliminary efforts along this line indicate that the Mi and Ni slopes are not too sensitive to the equipment capacity or the SEER rating. However the Oi slopes do depend on the SEER rating. It also

seems that the Q intercept is not sensitive to equipment capacity.

Table A3-2 provides a set of generic slopes and a Q-intercept for water-to-air cooling equipment. These values were generated by analyzing the performance data published by four different manufacturers (heat pump equipment that was manufactured between 1980 and 1986).

Generic Slopes and Intercepts Water-to-Air Cooling		
M Slopes	**N Slopes**	**O Slopes**
M1 = 3.33	N1 = 0.0002	O1 = 0.20
M2 = 500	N2 = -0.0475	O2 = 27.5
M3 = -110	N3 = 0.0325	O3 = 12.5
	N4 = 0.0010	
Q Intercept = 0.99		

Table A3-2

A3-5 Example — Water-to-Air Extrapolation

A model that approximates the performance for a particular water-to-air equipment package can be generated from the ARI certification data if a set of generic slopes (Mi, Ni and Oi) and a generic Q intercept are available. For example, the values that are summarized by Table A3-2 will be applied to the following ARI rating data:

• TC = 37,400 BTUH
• CFM = 1350
• Watts = 3840
• EWT = 70 °F
• EWB = 67 °F
• EDB = 80 °F

Step 1 - Evaluate the K and R Intercepts

$$37400 = K + (3.33 \times 1350) + (500 \times 67) - (110 \times 70)$$

$$K = 7105$$

$$3840 = R + (0.20 \times 1350) + (27.5 \times 67) + (12.5 \times 70)$$

$$R = 852$$

Step 2 - Write the Performance Equations

When the performance equations are assembled, they will have the following form:

$$TC = 20780 + (3.33 \times CFM) + (500 \times EWB) - (110 \times EWT)$$

SHR = 0.82 + (0.0002 x CFM) - (0.0475 x EWB) +

(0.0325 x RAT) + (0.0001 x EWT)

Watts = -600 + (0.20 x CFM) + (27.5 x EWB) + (12.5 x EWT)

A3-6 Comments on Integrated Capacity Data

In regard to air-to-air heat pumps, the published integrated heating capacity data (defrost penalty included) corresponds to a specific outdoor condition (dry-bulb and wet-bulb temperature). But when the equipment is installed in a home, the equipment package may be subjected to an outdoor condition that is quite different than the test condition. This means that the defrost penalty that is associated with the data published by equipment manufacturers will not be representative of the on-site defrost penalty.

The defrost penalty also depends on the defrost control hardware. For example, a system equipped with a demand defrost control will have a small defrost penalty when the climate is very cold or very dry and a large defrost penalty when the climate is moderately cold and wet. This occurs because fewer defrost cycles will be initiated when the coil icing potential is minimized.

Differences in integrated heating performance also will be noticed if the system is equipped with a simple defrost cycle timer. In this case the annual number of defrost cycles will not vary because the defrost cycle is initiated by a timer, but the average duration of the cycle will vary from climate to climate because the icing potential depends on the amount of moisture in the outdoor air.

A3-7 Air-to-Air Heating Model

The steady-state heating performance data of an air-to-air heat pump is a nonlinear function of the outdoor air temperature (OAT). This nonlinearity is associated with the defrost knee in the heating performance curve. A model of this heating performance data can be generated by using the following equations, which represent the heating capacity (BTUH) and the compression-cycle power requirements (Watts).

Heating BTUH = Ai + Mi x OAT

Watts = Bi + Ni x OAT

In this case, there are three sets of intercepts (Ai and Bi) and slopes (Mi and Ni). One set models the performance when the outdoor temperature is below 17 °F and above 47 °F. The second set models the performance when the outdoor temperature is between 17 °F and 35 °F. And the third set models the performance when the outdoor temperature is between 35 °F and 47 °F.

The following formulas can be used to evaluate the slopes and intercepts that are associated with each temperature range. To use these formulas, the ARI high (BTUH capacity at 47 °F) and low (BTUH capacity at 17 °F) heating capacity data and will be required, the ARI high and low power draw (Watts at 47 °F and Watts at 17 °F) will be required and the ARI defrost test output and input data (BTUH capacity at 35 °F and Watts at 35 °F) will be required.

Set 1 - Below 17 °F or above 47 °F

M = (BTUH @ 47 °F - BTUH @ 17 °F) / 30

A = BTUH @ 17 °F - (M x 17)

N = (Watts @ 47 °F - Watts @ 17 °F) / 30

B = Watts @ 17 °F - (N x 17)

Set 2 - Between 17 °F and 35 °F

M = (BTUH @ 35 °F - BTUH @ 17 °F) / 18

A = BTUH @ 35 °F - (M x 35)

N = (Watts @ 35 °F - Watts @ 17 °F) / 18

B = Watts @ 35 °F - (N x 35)

Set 3 - Between 35 °F and 47 °F

M = (BTUH @ 47 °F - BTUH @ 35 °F) / 12

A = BTUH @ 35 °F - (M x 35)

N = (Watts @ 47 °F - Watts @ 35 °F) / 12

B = Watts @ 35 °F - (N x 35)

A3-8 Water-to-Air Model

Since water-to-air equipment is not subject to a defrost cycle, the heating capacity and heating power requirement can be modeled with liner equations that are a function of entering water temperature (EWT). These equations are provided below.

Heating BTUH = A + M x EWT

Watts = B + N x EWT

In this case, one set of intercepts (A and B) and slopes (M and N) is adequate. The following formulas can be used to generate values for A, B, M and N.

$$M = (BTUH \text{ @ } 70\,^\circ F - BTUH \text{ @ } 50\,^\circ F) / 20$$

$$A = BTUH \text{ @ } 50\,^\circ F - (M \times 50)$$

$$N = (Watts \text{ @ } 70\,^\circ F - Watts \text{ @ } 50\,^\circ F) / 20$$

$$B = Watts \text{ @ } 50\,^\circ F - (N \times 50)$$

APPENDIX 4
Furnace Cycling Efficiency

A4-1 Cycling Efficiency

The cycling efficiency of a fossil fuel furnace or boiler depends on the percentage of the time that the furnace operates. At full load, the furnace will operate continuously and the cycling efficiency will be equal to the steady-state efficiency. At part-load conditions, the cycling efficiency will be less than the steady-state efficiency.

A4-2 Part-load Efficiency Curves

Figure A4-1 provides a set of curves that show the correlation between the cycling efficiency ratio (CER) and the heating load ratio (HLR). (The heating load ratio is equal to the heating load divided by the output capacity of the furnace.) These curves are based on research that was published by the Department of Energy and the National Bureau of Standards.

A4-3 Part-Load Efficiency Equations

The part-load efficiency curves that appear in Figure A4-1 can be represented by the following equations. In the third equation, C_i is a constant that has different values, depending on the AFUE rating that is associated with the furnace. On the next page, Table A4-1 provides a correlation between the furnace AFUE and the value of the C_i constant.

$$CER = \frac{Cycling\ Efficiency}{Steady\text{-}State\ Efficiency}$$

$$HLR = \frac{Heating\ Load}{Output\ Capacity}$$

$$CER = 1 - \left[\left(\frac{1}{e^{HLR}} \right) \right]^{C_i}$$

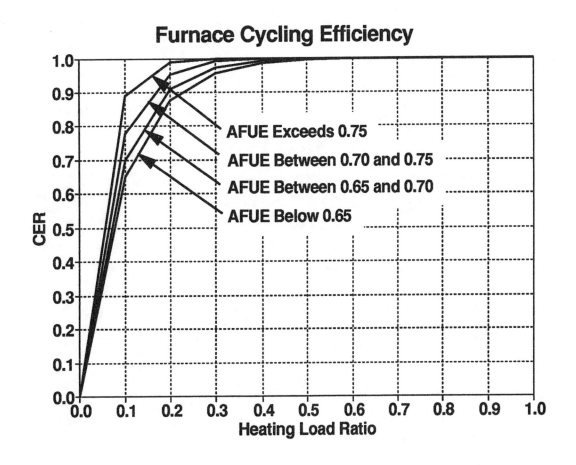

Figure A4-1

AFUE	C_i
Above 0.75	22.0
0.70 to 0.75	15.0
0.65 to 0.70	12.0
Below 0.65	10.5

Table A4-1

A4-4 Energy Consumption and Furnace Oversizing

The cycling efficiency curves (see Figure A4-1) explain why furnace and boiler operating costs are not sensitive to oversizing. Notice that the cycling efficiency ratio is essentially equal to the steady-state efficiency ratio when the heating load exceeds 25 percent of the output capacity. Also notice that a small reduction in the cycling efficiency ratio is associated with heating loads that are between 15 and 25 percent of the output capacity. As far as the annual heating-cycle energy requirement is concerned, the effect of the efficiency reduction in this range of load ratios is moderated by the small number of operating hours that are associated with this range of load ratios. Below 15 percent output capacity, the cycling penalty becomes very severe, but the net effect on the annual energy heating-cycle requirement is negligible because there are only a few hours of operation associated with this range of load ratios.

APPENDIX 5
DX Coil Matching

A5-1 Matching Evaporators and Condensing Units

The evaporator coil must be matched to the condensing unit to achieve the desired cooling performance. This is not a concern if cooling is provided by packaged equipment because the components are matched by the equipment manufacturer. However, contractors who mix and match evaporator coils and condensing units must select components that will provide the desired performance.

A5-2 Evaporator Performance

Evaporator capacity decreases as the suction temperature increases. This occurs because evaporator capacity depends on the temperature difference between the air that flows through the coil and the refrigerant that is evaporating in the coil. Also note that as the suction temperature increases, the surface temperature of the coil increases, causing the latent capacity of the coil to decrease.

A5-3 Condensing Unit Performance

The capacity of the condensing unit increases when the suction temperature increases. This occurs because the increase in gas density associated with a higher suction pressure translates into an increase in the amount of gas that is circulated by the compressor.

A5-4 Refrigerant-side Operating Point

The condensing unit and the evaporator will always balance out at a specific operating point. This operating point can be determined by plotting the condensing unit performance and the evaporator performance on a suction temperature versus total capacity chart. An example of this type of graph is provided by Figure A5-1. Notice that the components balance out at a specific tonnage and at a specific suction temperature. It is important to remember that the latent capacity of the coil is closely related to the suction temperature. If the balance point results in a suction temperature that is too high, the coil will not be able to control the humidity of the air that is circulating through the coil.

A5-5 Optimum Refrigerant-side Balance Point

Figure A5-2 shows that more tonnage can be extracted from a condensing unit when it is matched with a larger evaporator coil. Unfortunately the tonnage increase is accompanied by an increase in suction temperature and a corresponding decrease in latent capacity. This means that the indoor humidity could be unacceptably high if the evaporator is too large for the condensing unit. (Humidity control requires suction temperatures that cause the air off of the evaporator to have a temperature that falls between 54 °F and 58 °F db, depending on the latent load.)

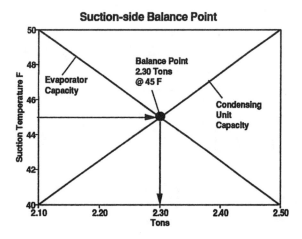

Figure A5-1

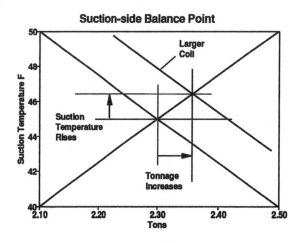

Figure A5-2

Appendix 6
Effect of Altitude

A6-1 Equal Mass Flow

Blowers are constant-volume devices. This means that the blower CFM is not affected by changes in altitude. In other words, a blower that moves 1,200 CFM at sea level will move 1,200 CFM in Denver. However, this does not mean that the sea level mass flow will be equal to the mass flow in Denver.

Since mass flow depends on the specific weight (Lb/Cu.Ft.) of the air that is being circulated by the blower, the mass-flow will decrease as the altitude increases because the density of the air decreases as the altitude increases. This is important because the capacity of an indoor refrigerant coil or furnace heat exchanger is affected by changes in mass flow. In other words, the output capacity (cooling or heating) at altitude will be equal to the output capacity at sea level if the mass flow at altitude equals the mass flow at sea level.

Since a manufacturer's performance data is associated with a sea-level installation, a flow rate adjustment is required if the equipment will be installed in a home that is located at a higher altitude. The density ratio values that are listed in Table A6-1 can be used to make this adjustment. Just divide the sea level CFM value by the density ratio to find the blower CFM that will make the "at altitude" capacity equal to the sea level capacity.

$$CFM \ at \ Altitude = \frac{Sea\text{-}level \ Flow \ Rate}{Density \ Ratio}$$

For example, if a piece of equipment is scheduled to be installed in Denver (about 5,000 feet), calculate the blower CFM that will be required to match the sea-level capacity rating. For this example, assume that the sea-level air flow rate is equal to 1,400 CFM.

$$CFM \ at \ Altitude = \frac{1400}{0.832} = 1680 \ (rounded)$$

Air Density Correction			
Altitude	Density Ratio	Altitude	Density Ratio
1,000	0.964	6,000	0.801
2,000	0.930	7,000	0.772
3,000	0.896	8,000	0.743
4,000	0.864	9,000	0.715
5,000	0.832	10,000	0.687

Table A6-1

A6-2 Derate Air-cooled Condensing Equipment

When altitude causes a reduction in the capacity of a forced convection heat transfer device, the lost capacity can be recovered by increasing the air-side flow rate (see the previous example). However, this may not always be possible. For example, an outdoor condensing unit may only operate at one fan speed. When this is the case, the capacity reduction will not be recoverable. In this case, the generic values listed in Table A6-2 can be used to derate the equipment. For example, if a single-speed condenser is rated at 36,000 BTUH at sea level, it will have 34,560 BTUH of capacity (0.96 x 36,000) at 5,000 feet.

Capacity Correction — Air-cooled Condenser				
2,000	4,000	6,000	8,000	10,000
0.98	0.97	0.95	0.92	0.90

Table A6-2

A6-3 Combined Penalty with Blower Adjustment

At altitude, the capacity of the indoor coil and the capacity of the outdoor coil will be reduced. However, the capacity of the indoor coil can be fully restored by increasing the blower speed (see Section A6-1). In this case, the reduction in the capacity of the complete package (indoor coil and outdoor unit) will be equal to the reduction in the capacity of the outdoor equipment (see Section A6-2).

A6-4 Combined Penalty — No Blower Adjustments

In some cases, the indoor blower may not be able to provide the extra margin of air flow that is required to compensate for an altitude-related capacity loss. When this is the case, the entire equipment package must be derated. The generic values that are listed in Table A6-3 can be used for this calculation. For example, if an equipment package is rated at 36,000 BTUH at sea level, it will have 33,660 BTUH of capacity (0.935 x 36,000) at 5,000 feet.

Capacity Correction — Air-cooled Package				
2,000	4,000	6,000	8,000	10,000
0.98	0.97	0.95	0.92	0.90

Table A6-3

A6-5 Furnaces

Furnaces that are certified by the American Gas Association (AGA) must be derated when they are installed at an elevation that is more than 2,000 feet above sea level. The generic values that are listed in Table A6-4 can be used for this calculation.

Output Capacity Correction — Gas Furnaces				
Altitude in Feet				
2,000	4,000	6,000	8,000	10,000
0.92	0.84	0.76	0.68	0.60

Table A6-4

For example, a furnace that has an output capacity of 48,000 BTUH at sea level will have 38,400 BTUH of capacity at 5,000 feet.

Output capacity = 0.80 x 48000 = 38400 BTUH

A6-5 Consult Manufacturer

Tables A6-2, A6-3 and A6-4 provide generic altitude correction values for air-to-air cooling equipment and furnaces. The values that are associated with a specific air conditioning unit, heat pump or furnace may be included in the performance data that is published by the equipment manufacturer. When this is the case, use the information that is provided by the manufacturer.

Appendix 7
Air-source Heat Pump Supply Temperatures

A7-1 Temperature Fluctuations

During heating, the temperature of the supply air that is delivered by an air-to-air heat pump gets noticeably cooler as the outdoor temperature declines. This means that there may be many operating hours associated with supply air temperatures that are 85 °F or colder. In fact, when the weather is extremely raw, the temperature of the supply air can fall below 80 °F. However, when the supplemental heat is energized, the supply air temperature may range from 90 °F to 120 °F or more, depending on the outdoor temperature and on how much supplemental heat is activated. So, when it is cold outside, the supply air temperature can be expected to fluctuate by 20 °F to 30 °F, or more.

Of course, the warmer supply air temperatures that are produced when the supplemental heat is energized are very desirable, however, this operating condition may only last for a few minutes out of each hour. Therefore, it is important to understand that an excessive amount of supplemental heating capacity produces short periods of operation with warm supply temperatures and long periods of operation with cool supply temperatures.

A7-2 Balance-point Diagram

Figure A7-1 shows the balance-point diagram for a home that is located in a cold climate. The heating load for this home is about 42,000 BTUH at an outdoor design temperature of 0 °F. Notice that the balance point is equal to 30 °F (BP1) when the heat pump is the only source of heat. When 5 Kw of supplemental heat is added to the system, the balance point is about 11 °F (BP2). When 10 Kw of supplemental heat is added to the system, the balance point is about -6 °F (BP3). And, when 15 Kw of supplemental heat is added to the system, the

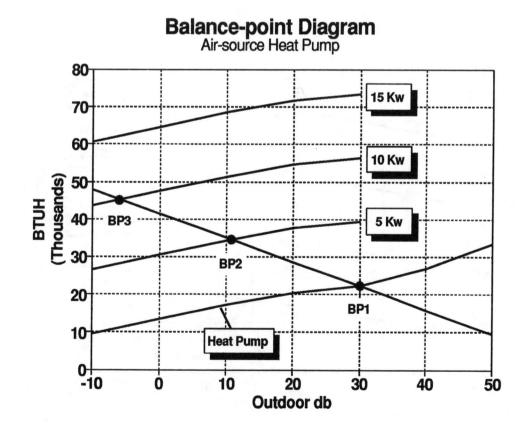

Balance-point Diagram
Air-source Heat Pump

Figure A7-1

balance point is off the chart, somewhere below -10 °F. (In this example, 10 Kw is an appropriate amount of supplemental heat because BP3 is just a little below the 0 °F design temperature.)

A7-3 Supply Air Temperatures

Figure A7-2 shows what the supply air temperature (SAT) would be when the heat pump is the only source of heat and when 5, 10 and 15 Kw of supplemental heat is energized. These supply air temperatures were estimated by using the sensible heat equation (below) to evaluate the temperature rise across the indoor air handler.

$$Rise\ (^{o}F) = \frac{Heat\ Pump\ BTUH + Supplemental\ BTUH}{1.1 \times Blower\ CFM}$$

$$SAT\ (^{o}F) = 70 + Rise$$

Figure A7-2 indicates that when no supplemental heat is energized, the supply air temperature will range from a low of 80 °F, when the outdoor temperature is 0 °F, to a high of 87 °F, when the outdoor temperature is equal to the 30 °F balance point. With 5 Kw of supplemental heat on line, the diagram shows that the supply air temperature will range from 93 °F to 100 °F. When 10 Kw is activated, the supply air temperature ranges between 106 °F and 113 °F. And, when 15 Kw energized, the supply air temperature ranges between 119 °F and 126 °F.

A7-4 Supplemental Kw Run Fraction

As indicated by Figure A7-2, a larger amount of supplemental heating capacity will produce a warmer supply temperature. However, this condition will only occur for a few minutes out of each hour. This means that there will be many minutes per hour associated with the "heat pump only" supply temperature, which is relatively cool. This behavior can be quantified by using the following equation, which evaluates the number of minutes out of every hour that the supplemental heat will be energized.

$$Min/Hr = 60 \times \frac{Heating\ Load - Heat\ Pump\ Capacity}{3413 \times Supplemental\ Kw}$$

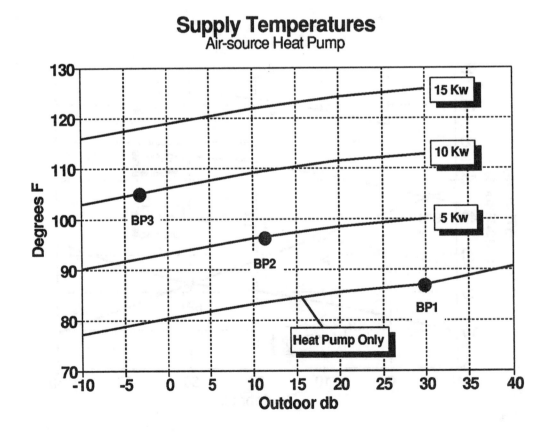

Figure A7-2

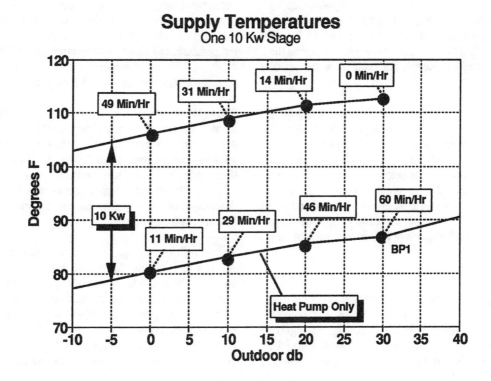

Figure A7-3

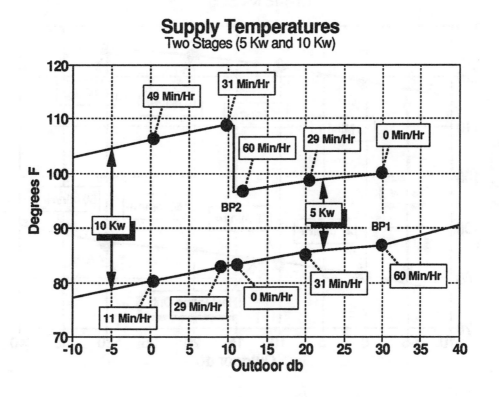

Figure A7-4

On the previous page, Figures A7-3 and A7-4 show the supply air temperatures and the corresponding run fractions for air-to-air heat pump packages that have 10 Kw and 15 Kw of supplemental heat. Notice how the 10 Kw system causes the supply temperature to be warmer for more minutes out of every hour.

For example, at the 0 °F outdoor temperature, the supply temperature of the 10 Kw system is above 100 °F for 49 minutes out of every hour compared to 33 minutes per hour for the 15 Kw system. It follows that the 10 Kw system will have a supply temperature that is below 85 °F for 11 minutes out of every hour compared to 27 minutes per hour for the 15 Kw system. A similar analysis at other outdoor temperatures indicates that the 10 Kw system will always provide more minutes per hour of 100 °F plus supply temperatures than the 15 Kw system.

A7-5 Two Stages Improve Performance

When the outdoor temperature is above 10 degrees Figures A7-3 and A7-4 indicate that both the 10 Kw system and the 15 Kw system spend most of the hour operating at the lower, "heat pump only" supply air temperature. Therefore, during moderate weather, the 10 Kw system is not that much better than the 15 Kw system. For example, at the 20 °F outdoor temperature, the supply temperature of the 10 Kw system is above 100 °F for 14 minutes out of every hour compared to 10 minutes per hour for the 15 Kw system.

The performance of the 10 Kw system can be improved by using two 5 Kw coils instead of one 10 Kw coil. Note that two outdoor thermostats will be required for this design. When the outdoor temperature drops below 30 °F, the first outdoor thermostat can be used to energize 5 Kw of supplemental heat, and when the outdoor temperature drops below 11 degrees, the second outdoor thermostat can energize an additional 5 Kw of supplemental heat.

The benefit of a staged system is demonstrated by Figure A7-5, which shows that when the outdoor temperature is above 11 °F, one 5 Kw stage of supplemental heat results in warmer supply air temperatures for more minutes out of every hour. For example, at 12 °F, the supply temperature of the staged system is above 95 °F for about 60 minutes out of every hour, compared to 31 minutes for the 10 Kw system. And, at 20 °F, the supply temperature of the staged system is above

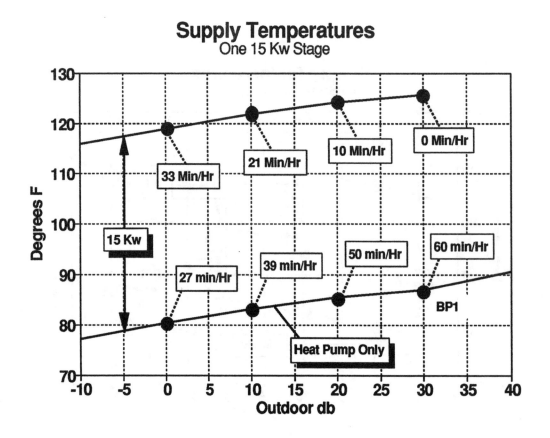

Figure A7-5

98 °F for about 29 minutes out of every hour, compared to 14 minutes for the 10 Kw system.

A7-6 Sizing Supplemental Heat

Keeping the supplemental Kw to a minimum will reduce the number of minutes per hour that the system will operate with relatively cool supply air temperatures. The maximum amount of electric resistance heat that is controlled by the second stage of the room thermostat should not be much larger than the supplemental heat requirement that is associated with the balance point diagram. (The supplemental heat requirement is equal to the difference between the heat that is required on a winter design day minus the heat pump output when it operates at the winter design temperature.) This means that if an additional increment of electrical resistance heat is required to satisfy a 100 percent emergency heat requirement, it should be controlled by a manual switch.

A7-7 Staging Supplemental Heat

Staging the supplemental heat will increase the number of minutes per hour that the system will operate with a relatively warm supply temperature. As indicated above, two 5 Kw stages provide warmer supply temperatures when the outdoor temperature is between the first balance point and the second balance point (10 °F and 30 °F in this case.) This range of outdoor temperatures is very important because most of the winter heating hours are associated with the intermediate temperatures.

Index

Notes

Notes

Notes

Notes